Andrei Moldavanov

Evolutionary Aspects of Energy Exchange in Open Systems

Physics of Genetic Code and *DNA*

DOI: https://doi.org/10.52305/RIRN1446

Library of Congress Cataloging-in-Publication Data

ISBN: 979-8-89530-389-4 (Softcover)
ISBN: 979-8-89530-614-7 (e-Book)

Published by Nova Science Publishers, Inc. † New York

Contents

Preface

Broadly, this book represents an alternative look at the theory of stochastic exchange in an open system. More technically, it studies an energy exchange in a thermodynamic system at the condition of the highest possible degree of system openness in mathematical terms.

So far, the author has tried to apply this theory in a few fields, such as natural selection, general data science, the development of magnetospheric disturbances, non-equilibrium thermodynamics, and some others. As the response of the auditory was positive rather than negative, this encouraged the author to think about, on the one hand, the further popularization of developed theory and, on the other hand, the expansion of it to not yet covered fields, first of all, to genetics and evolutionary biology. So, in this book, the recent findings are analyzed from the perspective of the linkage from, to some extent, abstract concepts in the arrangement of energy exchange to fundamental physical features in the structure of genetic code (*GC*) and deoxyribonucleic acid (*DNA*). Here the idea of publishing a new book was born.

The book's main objective is to introduce in a unified manner the major principles of energy development of a system under the condition of full openness and compare the obtained results with actual signatures of natural processes. Thereby, the author sees his mission in helping the reader to develop a working knowledge of high-level evolution mechanisms, i.e., to develop the skills and background needed to recognize, formulate, and solve problems of growth using the suggested toolset.

In general, *Evolutionary Aspects* is a succinct and easy-to-follow book for physical and mathematical grounds in the syllabus of evolutionary biology. This book is meant for the researcher, scientist, or engineer who uses the concept of open systems, in the wider sense, anyone interested in the application of evolutionary mechanisms. This includes, naturally, those working directly in the field of theoretical biology and genomics, as well as many others who use an evolutionary approach in fields like economics, finance, and many fields of science and engineering. The only background required from the reader is some knowledge of Euclidean geometry, calculus, linear algebra, thermodynamics, basic probability theory, and combinatorics. Prior exposure to *GC* and *DNA* fundamentals is desired on a surface level, generally, without chemistry and biology details, which ultimately could be skipped. Eventually, the only skill a reader must have is common sense. If it is, he or she should be able to follow every significant argument and discussion in the book. After all, this book is a guideline on how to track the natural way for things to grow, so the author hopes that readers can still get all of the essential ideas and important points.

To assist the reader, the author has given short conclusions at the ends of all chapters. Though probably a large number of the points demand a more comprehensive treatment.

The book is divided into three parts. The chapters of the first part are primarily intended to provide an overview of the applying techniques; the second part deals with the evolutionary aspects of stochastic exchange in open system, and part three is about applications of the findings of the previous parts in the construction of the numerical basics of *GC* and *DNA*.

Chapter 1.1 provides a brief course in the substantiation of the taken approach and dedicated language. In Chapter 1.2, the spectrum of energy exchange in *OTS* is presented; in Chapter 1.3, verification of the used approach is conducted. Then, chapter 2.1 considers a morphology of a stochastic energy exchange; chapter 2.2 looks at bifurcation properties of the found spectral nodes; and chapter 2.3 analyses energy exchange in the assumption of the existence of time symmetry breaking. In chapter 3.1, a mathematical analysis of origin for genetic code is suggested, while in chapter 3.2, further continuance to the area of *DNA* is discussed. In appendixes, the cumbersome geometrical and combinatorial relations semantically connected with body text are presented.

All in all, the aim of this book is to enable the reader to go beyond the ordinarily encountered standard understanding of the phenomenon of energy exchange. Each quantitative model certainly has its strengths and drawbacks, and it is better to have a larger set of available tools.

In conclusion, I would like to thank Nova Science Publishers for the opportunity to present my results on this topic. Any errors and omissions are entirely my responsibility.

Finally, I owe a debt of profound gratitude to my beloved wife, Elena, who remained a constant source of support and inspiration during the writing of this book. Also, I thank my sons, Sviatoslav and Vladislav, as well as Kate, for their vital lifestyle in maintaining a creative family atmosphere.

Andrei Moldavanov
Burnaby, British Columbia

Introduction

It is well known that a commonly accepted physical and mathematical platform for natural evolution is not built [1, 2], while researchers do not mind that physical constraints can and even must influence both development understood as forming of the cognitive behavioral changes and evolution [14]. In this context, evolution can be understood as a gradual process of change over time in the population of organisms present in identical conditions [2]. Notice that mentioning the time dependence looks quite reasonable since time is usually considered a universal characteristic in any natural process.

However, we may also think about evolution in respect of some different quantity (quantities), which can depend on time. Of course, the quantity that has the privilege to describe such a majestic and many-sided process as evolution should be sufficiently special.

If we take a look at the entire history of natural sciences, then a unique association between conceptions of time and energy is being immediately observed [3]. Though time is not directly related to energy itself (some researchers believe otherwise [4]), it is definitely related to many facets of energy.

So, this book is not about the time sweep of evolution; instead, evolution is considered as a the consequence of changes over energy. Saying "energy," we mean not only a physical quantity measured in "joules" and characterizing the quantitative ability of a body or a physical system to perform some action.

Instead, it assumes all sorts of physical quantities, which, from their angles, describe changes in the action, current or postponed, status of the body (system) regarding the existing environment. Momentum, charge, mass, and some others are in this list, as discussed below.

Of course, after all, energy and its co-values change over time, and ultimately it is possible to say that our research is reduced to an investigation of changes in time anyway. To remove possible misunderstanding, in this book we do not use the time derivatives d/dt and $\partial/\partial t$ of any order. Doing that, in the meantime, we allow energy in the background to depend on time, however, our results formally do not directly call for time variations.

So, resuming what has been said so far, it looks justified to think that study of natural evolution over energy or some energy-related quantity is physically reasonable and can be considered as a useful alternative to conventional time evolution [12, 13].

This book was not planned to be used as a reference for mechanisms of chemical, biochemical, or biological transformations that may or may not accompany physical alterations. Though we are completely aware that described physical results have to include significant or even indispensable chemical and biological context, the latter are considered in the quite concise fashion of exposition. Readers may discover suitable references on their own; at these times, it is not a big challenge. We will keep focusing on the physics of matter.

So, since the mid-nineteenth century, the Darwin-Wallace mechanism of natural selection has been the main driving force for the origin of new species, and this view, with some modifications, has survived to the present day [15].

Under the influence of Turing's model of morphogenesis [9], physics and mathematics started on exploration of the before not achievable biology fields [22]. As of today, it is well realized that physical factors may play a major role in early evolutionary history through different energy channels.

From physical grounds, any process, and evolution is not an exception, needs some form of energy support to move forward. Rephrasing, it presumes that the permanent energy flow between open thermodynamic system (*OTS*) and external environment is a critical condition of evolution. Ultimately, these flows may reveal evolutionary potential of *OTS* through a gradual complication of its structure and functions.

The process for establishing links *OTS* ↔ surroundings can be described in the language of conservation laws without any need to consider microscopic details of the system [24].

It is a proven fact that the physico-chemical conditions on the Earth since its segregation as a separate body have been changed radically. However, it did not stop the evolutionary process. In the author's opinion, it says that the driving force of evolution is an almost universal mechanism that can work in literally any physical environment.

In the description of evolution as any other complex natural phenomenon, we have problem with too many unknowns. In this situation, the generic system approach could be one of the possible solutions. Besides, the very universal and simple physical mechanisms should be accounted for accurate simulation of conditions on early Earth.

It is well known that complex dynamical systems transform energy from the environment in maintaining themselves at an energy distance from equilibrium and can hold energy in non-linear relationships among system components supporting the self-amplifying [14].

Besides, as exact energetics on young Earth that could convoy origin and development of original forms is unknown [23], it is quite possible that earliest forms, through their membrane transductions, used as many energy sources as possible, such as light, thermal energy, pressure, touch, stretch, movement of water and air, gravity, electric and magnetic fields, as well as a wide range of chemical substances [24, 14].

The keystone mathematical hypothesis of our research is that the major discriminative difference between simpler and more complicated things lies in a composition of established linkages between entities of interest and their external environment, which this entity is capable of serving. In other words, the number and quality of bidirectional links of different essences comes as an essential discrimination parameter.

The next step is to find a carrier for physical implementation of the above linkage. By default, a concept of linkage presumes the presence of suitable transportation processes, meaning transport of energy, matter, and so on. Fortunately, in physics, the concept of similar carrier has been known for decades. It is an *OTS* [23]. In contrast to other system models such as isolated (no connection with externals at all) and closed (more or less serious restrictions in external connections), *OTS* has full freedom for communication, including the non-equilibrium processes.

Summarizing above, we can certify that despite significant progress, adequate description of *OTS* evolution based on strict use of the known theoretical baggage like physical and mathematical laws remains one of the main challenges of modern natural science and technical progress [15-21]. At this, the interaction of energy fluxes in a suggested model of *OTS* can take different forms, meaning exchange between energy quantities targeting physical, chemical, biological, and informational processes [25], which warrants the development of a theory for such interaction.

So, the main goal of this book is to consider the phenomenon of unforced evolution using the most universal laws of our physical world and show remarkable logic and rationality of internal evolutionary operations, starting from quantity relations managing the appearance of evolutionary infrastructure of exchange to the details of assembly and geometry forms of *DNA*.

Hopefully, the reader finds the suggested narrative useful.

Part 1.
The Basics of Stochastic Energy Exchange in *OTS*

Chapter 1.1

The Model of Stochastic Energy Exchange (*RECE)*

1.1.1. Substantiation of Approach (Basics)

Questions raised in the introduction appeal to the evolution of open systems and the concept of *OTS*. We believe that one of the possible answers may include some universal quantity that is naturally integrated with both *OTS* and flows of different nature between *OTS* and the external world. To be more specific, we restrict our consideration to only such *OTS*-externals relations that obey the conserved laws and their differential form, known as the continuity equation (*CE*).

Indeed, the platform of conserving law and *CE* looks sufficiently convenient. Firstly, *CE* is of actual universal relation, which works successfully in different physical conditions. The physical scenarios when *CE* does not work are quite uncommon [37]. Secondly, *CE,* by default, serves as an interface between *OTS* and external energy flow. Thirdly, the mathematical structure of *CE* is simple to avoid cumbersome calculations and, where practicable, to obtain exact solutions.

We will use standard probabilistic methodology for the description of random values, for example [28, 36]. In this case, the concept of phase space is predominant. The usual spatially-time method will be used for the description of deterministic quantities.

So, the cornerstone of this book is an idea of the conserved link between *OTS* and its externals. From a physical standpoint, the formalism of conserved links works for quantities that meet the criterion of conservation (1.1).

Traditionally, the category of conserved quantities (further *CQ*) contains energy, electric charge, momentum, angular momentum, and mass, as well as some other physical quantities from the world of elementary particles. To underline the different physical essences of *CQ*, we will call this the charge-energy-mass (*CEM*) approach. The motivation is to highlight that all *CQ*, mathematically, can be formalized in an identical way (1.1, 1.2).

By definition, *CQ* obeys the appropriate *CE,* so mathematically, a single *CQ* connection in non-relativistic approximation in $R \text{->} R^3$ and J: $R \rightarrow R^+$ can be written in the form:

$$\frac{\partial \varepsilon}{\partial t} + \nabla \cdot \mathbf{J} = 0 \tag{1.1}$$

where J is the flow of some *CQ*, ε is its volume density, t is time, ∇ is Nabla operator.

Note that as information does not have the property of conservation, it is not accounted for to contribute to (1.1, 1.2).

Following the above logic, the presented approach incorporates those exchange scenarios that necessarily have an energy dimension. In other words, we account for the exchange scenarios in which change in the system state is projected onto an energy axis for sure, no matter what sort of exchange process or combination of processes (thermal, chemical, optical, mechanical, or any other) it deals with. It is quite a natural step for *CQ* to have an energy measure, for example, the well-known energy equivalent of mass [38, 39].

To support the above approach, we have to make sure that a suitable energy continuity equation (*ECE*) that covers any possible act of exchange exists. The guarantee that such *ECE* is found without fail is the infiniteness of the set of *ECE,* which is generated by the joined work of all implicated *CQ*. So, we come to the idea of an infinite number of *ECE* (energy links) given by (1.1). Obviously, although this is not a requirement, each *CQ* can have its own infinite system of suitable continuity equations, likewise (1.1).

Formally, at every single moment, all links are available for energy transfer, and the required combination of links is taken at random. At this point, some links can become instant sources (transfer inwards), while others become instant sinks (transfer outwards). So, *CEs* for all non-energy *CQ* are accounted for and contribute to the top-level *ECE*.

To include mass flow, an assumption of the permeable boundary of the *CE* system is used. Also, let our system be in a permanent energy exchange with a thermal bath of unlimited capacity.

In accordance with the above, flow J denotes a multitude of suitable flows of various essences, external and internal, that influence energy. In a similar way, volume density ε is treated as a local aggregate of the properties of the area that are responsible for the storage and utilization of energy.

It is specifically provided that in this formulation of a problem, an algebraic sum of energy flows is not considered at all. It is quite the contrary; at each moment of time, the random choice of an individual link and the direction of flow in this link take place. In accordance with that, it is not possible for the zero flows J_{in} and J_{out} to coexist, where J_{in} means flow inwards *OTS*, and J_{out} signifies flow outwards *OTS*. At this point, any intercoupling between single links is prohibited.

Pay attention to the fact that, from all possible probability distributions, the maximum entropy for a limited continuous random quantity without any restrictions but associated with its range is achieved for the uniform distribution *U(a,b)*, where *a* and *b* are the lower and upper borders of the range of change, respectively [40]. Based on that, for further description of random processes, we will be using a unified probability distribution.

1.1.2. Randomized Energy Continuity Equation (*RECE*)

Based on the above, the mathematical model of *OTS* can be written as an infinite system of *CE* (1.1).

$$\begin{cases} \frac{\partial \varepsilon_1}{\partial t} + \nabla \cdot \mathbf{J}_1 = 0 \\ \frac{\partial \varepsilon_2}{\partial t} + \nabla \cdot \mathbf{J}_2 = 0 \\ \dots \\ \frac{\partial \varepsilon_n}{\partial t} + \nabla \cdot \mathbf{J}_n = 0 \\ \dots \end{cases} \tag{1.2}$$

In (1.1), conduct an equivalent mathematical transformation. In the first step, get rid of dependence on *t* and spatial operator ∇.

So, let *OTS* be the closed manifold oriented by inward-pointing normal *n* at each point of the boundary *S*. Suppose the direction of dJ is determined by the unit vector *m*, $m = (\cos \varphi_x, \cos \varphi_y, \cos \varphi_z)$, where $\varphi = \angle(dJ, n)$,

$$\cos \varphi_i = \frac{J_i}{J}, \tag{1.3}$$

dJ_i is i-th component of dJ, dS is a shorthand for ndS, $i = \{x, y, z\}$ in the Cartesian system.

Multiply both sides of (1.1) by elementary volume dV and use (1.3), so we have

$$\frac{dU}{dt} = -dJ(\mathbf{m}\cdot d\mathbf{S}) = -dJ(\mathbf{m}\cdot\mathbf{n})\,dS \tag{1.4}$$

where $m\cdot n = cos\ \varphi$, $dU = \varepsilon\cdot dV$.

Divide (1.4) by $J\cdot dS$ and observe $Q = J\cdot dS\ dt$, so

$$\frac{dU}{Q} = -d\,(\ln y)\,x \tag{1.5}$$

where from this point on random $x = cos\ \varphi$, y (energy exchange rate) $= J/J_0$, normalizing constant $J_0 > 0$. Notice that the support range for random x is determined as $|2lny|$.

Note that by construction $\varphi \in R^3$, so$|\ x|\ \leq 1$. At variable x, it naturally accounts for an infinite n in system (1.2).

It is worth noting that (1.5) is not an exact differential, which is why the function dU/Q is not a function of state.

Equivalently, from (1.5)

$$\frac{dU}{\delta Q} = -x \tag{1.6.a}$$

where $\delta Q = dJ\cdot dS\ dt$.

Further, formalisms (1.5) and (1.6.a) are called randomized energy continuity equation (*RECE*).

Key role in (1.6.a) plays x in the right-hand part. In the trivial case, when we have only one equation (1.1) and, appropriately, $x = 1$, equation (1.6.a) becomes classic *ECE*

$$\frac{dU}{\delta Q} = 1, \tag{1.6.b}$$

declaring exact equality between quantity δQ brought to an interface *OTS* from storage reservoir and quantity dU, ultimately acquired by *OTS*, i.e., $dU \equiv \delta Q$.

However, if it is necessary to account for an infinite number of random links with reservoir (1.2), quantity δQ brought to an interface *OTS* and quantity dU factually acquired by *OTS* are not necessarily the same, $dU \neq \delta Q$. In this case, the non-trivial scenario (1.6.a) with $x \neq 1$ could be the case.

Note that whether δQ is fully converted to dU or not depends on specific individual conditions (random x) at interaction *OTS* and external environment. Then, ratio dU/Q can serve as an indicator for an instant state of bidirectional flow of energy, i.e., *an* indicator of the state of interface for *OTS*, which can be in equilibrium or a non-equilibrium state with the reservoir, and (1.6.a) could be considered a mathematical instrument for the description of the general process of exchange between *OTS* and reservoir through flow *J*.

For the purpose of this book, it is crucial to clearly separate the scope of (1.6.a) and (1.6.b). While (1.6.b) deals with the point of local contact *OTS* with flow in a mathematical sense, (1.6.a) does it for some finite (physical) vicinity of contact. In other words, by replacing (1.6.b) with (1.6.a), we do step into reality and, actually, replace the predetermined approach (1.6.b) with the probabilistic approach (1.6.a).

Mathematically, interaction between an energy link and *OTS* goes through formalism *CE*. So, the difference between deterministic *CE* (1.1) and stochastic *RECE* (1.6.a) is that *CE* deals with predefined transport through one fixed channel, while *RECE* treats interaction with an infinite number of channels, each time randomly selecting a path for energy transport.

Regarding the difference between (1.6.a) and (1.6.b), it is necessary to pay attention to one more principal moment. While case (1.6.b) can be used to describe equilibrium state, case (1.6.a) is closely related to the situation when permanent reproduction of multi-versioning of development occurs (this conclusion also will be confirmed in chapter 1.2). Hence, the greater number of development scenarios that exist, the validity of (1.6.a) is higher. Further, to describe the dynamics of the *OTS* interface, we will be using (1.6.a), which can be used for the non-equilibrium systems.

So, we substantiated that (1.6.a) is an accurate mathematical tool for coherent description of exchange *OTS* with storage reservoir. Abovesaid explains reasons to name (1.6.a) as *Randomized Energy Continuity Equation* (*RECE*).

Formally, *RECE* cannot be considered an equation of "continuity" any longer due to, generally, a mismatch between dU and δQ, but we are going to keep this name for consistency.

An infinite system (1.2) can be converted to equivalent stochastic equation (1.5) or (1.6.a), in which flux J is a deterministic quantity, whereas an interface factor x is random quantity in the range *[-1, 1]*.

1.1.3. Phase Space

Based on (1.5), introduce function

$$\Upsilon = \iint_D \left(\frac{dU}{Q}\right)_R = -\iint_D \frac{dy}{y} x \tag{1.7}$$

where $D \subseteq R^2$ is phase space for all possible states dU/Q.

Then, taking Riemann integral on dy, we obtain

$$\Upsilon = -\int x \int \frac{dy}{y} = -\int_D \delta\Upsilon(x, y), \tag{1.8.a}$$

where random

$$\delta\Upsilon(x, y) = x \ln y \tag{1.8.b}$$

Root $\{y = 1, x = 0\}$ of equation $\delta\Upsilon = 0$ is a point of stationarity of the sought solution, designate it $SP\ (P_0)$.

The non-integrable structure of (1.8.a) warrants usage of the phase (exchange) space. As x is a continuously distributed random quantity, then random

$$\delta\Upsilon = \delta\Upsilon(x \cdot \ln y) : \Phi \to \Delta\Upsilon \tag{1.9}$$

is a measurable function from the set of possible outcomes $x \cdot lny$ to some set $\delta\Upsilon$ with Φ as a probability space and $\delta\Upsilon$ is a measurable space.

Variable y has continuous deterministic representation in range $R \to R^+$, variable x changes in the range *[-1,1]*, therefore $\delta\Upsilon$ randomly takes values from continuous subspace of dimensionality M = [-1, 1] * lny, where symbol * signifies point-wise multiplication [2].

Then, $\delta\Upsilon$ is in the space between branches $-ln(y)$ and $ln(y)$, being limited from the left by $y = 0$. Further will be shown that $\delta\Upsilon$ has meaning only at $0 \leq y \leq e$, where e is Euler number (also called point GRP_R). Hence, a range of $\delta\Upsilon$ (call it G) varies from the range $(-\infty, \infty)$ at $y = 0$ to the one-point set $\delta\Upsilon = 0$ at $y = SP = 1$, and further to the range $[-1, 1]$ at $y = GRP_R$. Each point in G fits into one microstate of *OTS*. Any change in the microstate of *OTS* is equivalent to the movement of a given point along a branch of the solution, forming a system trajectory in phase space.

Branches $-ln(y)$ and $ln(y)$ seem to be saying the following. Branch $-ln(y)$ fits to positive changes $dy > 0$ (*OTS* receives some amount of energy), while branch $ln(y)$ fits to negative changes $dy < 0$ (*OTS* loses some amount of energy). It means that the processes of receiving and giving back energy are topologically separated. As a result, the real process of energy exchange (i.e., sporadic interleaving of phase receiving and return of energy) from a topological standpoint takes the form of closed random trajectories (loops), with examples in Figure 1.1 both for the pre-stationarity area $y < SP$ and the post-stationarity area with $y > SP$, where SP is the point of stationarity.

In this sense, points $y = 1/e$ (further GRP_L) and $y = e$ (further GRP_R) in considering model play the role of turning points when *OTS* falls in the domain of original/terminal chaos, and, as a result, any energy uniqueness of *OTS* is simply erased. The meaning of GRP_R, GRP_L, and other special points will be highlighted in chapter 1.3, when the morphology of *OTS* evolution will be discussed.

Note that the formation of a loop in the space of state density matrices agrees with the known ideas, for example [1], stating that the appearance of such a loop is a consequence of the non-trivial topology of the space with a path-dependent phase.

In Figure 1.1, two demonstrating loops are shown. In the range $0 \leq y \leq SP$ contour C_1 with clockwise navigation and in the range $SP \leq y \leq GRP_R$ contour C_2 with counterclockwise navigation. Introduce angle p between normal to navigation plane, directed from plane of figure to observer, and direction of movement such that clockwise

$$cos\ p = 1 \tag{1.10.a}$$

and counterclockwise

$$cos\ p = -1 \tag{1.10.b}$$

It is worth noting that opposite directions, i.e., counterclockwise in range $0 \leq y \leq SP$ and clockwise in range $SP \leq y \leq GRP_R$ are prohibited.

1.1.4. Solution for Energy Efficiency Υ

To (1.8.a) apply integration of pointwise product [2] and, assuming y unlimited in its range,

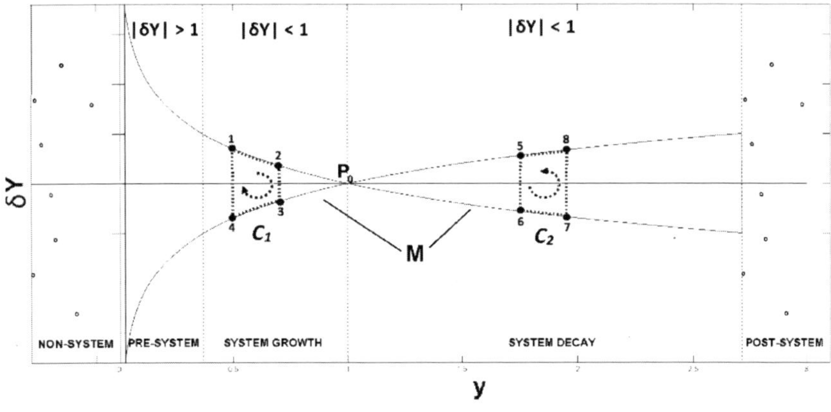

Figure 1.1. Phase space y - $\delta\Upsilon$ for randomized energy continuity equation (*RECE*). In the plot, by an abscissa axis, a unitless energy rate $y = J/J0$ is indicated, and by an ordinate axis, the instant efficiency of energy exchange $\delta\Upsilon$. Space M (marked in light grey) includes two subspaces, the left with $x \in [-1, 1], y \in [0, 1)$ and right with $x \in [-1, 1], y \in [1, GRP_R)$ separated by a point of stationarity P_0. In the plot, part (1) "Original absence of *OTS* (Non-System)" $(y < 0)$, part (2) "Forming of *OTS* (Pre-System)" $(0 < y < GRP_L)$, part (3) "Development of *OTS* (System Growth)" $(GRP_L < y < 1)$, part (4) "Reverse development of *OTS* (System Decay)" $(1 < y < GRP_R)$, part (5) "Terminal absence of *OTS* (Post-System)" $(y > GRP_R)$. At stage $(0 \leq y \leq GRP_L)$, quantity $|\delta\Upsilon| > 1$, while at $(GRP_L \leq y \leq GRP_R)$ quantity $|\delta\Upsilon| < 1$. At stage (1) *OTS* has not existed yet, at stage (5) *OTS* has not already existed, so quantity $\delta\Upsilon$ is not defined. Also, two contours, C_1 and C_2, with opposite sense of circumnavigation are shown.

$$\Upsilon(y) = -\int x * \ln y = -\int_{-1}^{1} dx \cdot \int_{0}^{y} \ln y \, dy = -by(\ln y - 1) \tag{1.11}$$

(Figure 1.2), where b is normalization factor.

Use condition

$$\Upsilon(y=1) = 1$$

then $b = 1$ and integral efficiency of energy exchange (rate of total energy exchange dT/dy).

$$\Upsilon(y) = y - y\ln y \tag{1.12.a}$$

or

$$\Upsilon(y) = k_X\, y \tag{1.12.b}$$

where

$$k_X = 1 - sign\,(\ln y)|\ln y| \tag{1.12.c}$$

and accounted that at $y > 1$ sign of $ln\, y$ changes. Also, total energy exchange

$$T(y) = \int_0^y \Upsilon(z)dz = \frac{3y^2}{4} - \frac{y^2}{2}\ln y = \frac{y^2}{4}(3 - 2\ln y) \tag{1.12.d}$$

Finally, note that $\delta\Upsilon/dy$ changes its sign in *SP* as at $y < 1$

$$\frac{\delta\Upsilon}{dy} \geq 0$$

while at $y > 1$ (1.12.e)

$$\frac{\delta\Upsilon}{dy} \leq 0$$

Note that (12.b,c) does not impose any restrictions on factor k_X, i.e., k_X is a continuous quantity.

Find y meeting condition $\Upsilon(y) = 0$. Then, resolving (1.12.a) with respect to y, obtain

$$\begin{cases} y_1 = \exp[1 + W_L(-1, -\frac{\Upsilon}{e})] = 0 \\ y_2 = \exp[1 + W_L(0, -\frac{\Upsilon}{e})] = e \end{cases} \tag{1.12.f}$$

where W_L is Lambert's function, $W_L(0)$ and $W_L(-1)$ its upper and lower branches, respectively [3]. In chapter 1.2, it is shown that $\Upsilon = 0$ holds not only at $y = GRP_L$ and $y = GRP_R$, but also in an infinite set of points within y-range (GRP_L, GRP_R).

Based on the found above borders $y = 0$ and $y = GRP_R$, further we limit our research only by the y-range $0 \le y \le GRP_R$. Those cases, when coming out the right border GRP_R will be specially discussed.

Note that M possesses an axial symmetry in respect of line $x = 0$ and the logarithmic symmetry with respect of the line $y = 1$. It means that for any non-zero x_0, $y_1 \in [0,1]$, $y_2 \in [1, \infty]$ holds

$$\int_{y_1}^{y_2} dy \int_{-x_0}^{x_0} \delta\Upsilon(x,y)\, dx = 0,$$

where $y_{k1} \cdot y_{k2} = 1$.

In chapter 1.3, we will be interested in only parts of subspace M_d, where $y_1 \in [GRP_L, SP]$, $y_2 \in [SP, GRP_R]$.

As for the whole M_d holds

$$\int_{GRP_L}^{GRP_R} dy \int_{-a}^{a} \delta\Upsilon(x,y)\, dx = 0,$$

then

$$\left| \int_{GRP_L}^{SP} dy \int_{-a}^{a} \delta\Upsilon(x,y)\, dx \right| = \left| \int_{SP}^{GRP_R} dy \int_{-a}^{a} \delta\Upsilon(x,y)\, dx \right| \tag{1.13}$$

where $0 \leq a \leq SP$. Ratio (1.13.b) means that the rate of energy production going through in the left submanifold exactly matches the rate of energy production going through in the right submanifold.

So, the major difference of current research from theories developed in [4] and [5], and similarity to the discussed in [6], consists in what the research object is not *OTS* itself, but a product of interaction between *OTS* and energy flow, i.e., interface of pair flow-*OTS*, that reflects the dynamics of their interaction and mutual changes that happen.

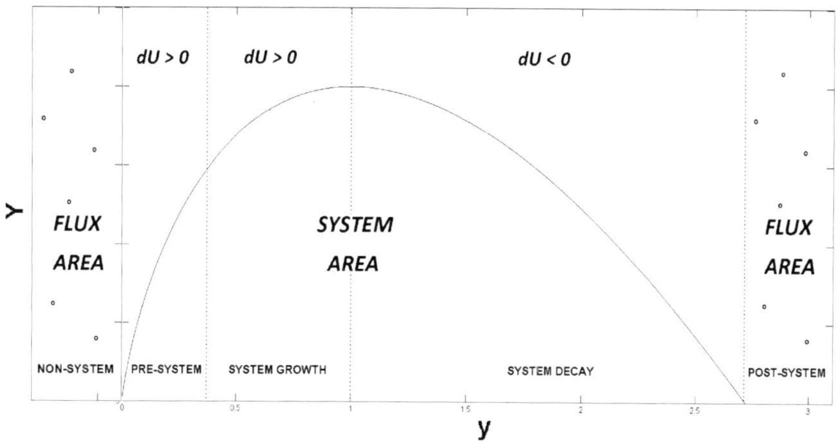

Figure 1.2. Dependence of average efficiency of energy exchange Υ on instant rate of energy exchange y. In the plot, the designation of an abscissa axis is the same as in Figure 1.1. By an ordinate axis the average efficiency of energy exchange Υ is indicated. Graph demonstrates the lifetime cycle of Υ at y in $[0, GRP_R]$. Two major areas "Flux Area" and "System Area" are shown. Subareas "Non-System" and "Post-System" are combined and denoted as "Flux Area" in contrast to three central areas ($0 \leq y \leq GRP_R$) that are combined and denoted as "System Area." In "Flux Area," there is a flux only, while in "System Area" there is a result of the flux-system interaction.

1.1.5. Entropy of *OTS* Energy Interface

In this model, *OTS'* dynamics obey *RECE* (1.6.a). At that, an additive structure of Υ (dependence on traversed path y and independence on the sign of dy) presumes accumulation of information on change in an interface state. In this sense, *OTS'* evolution is a non-Markovian process [7]. By comparing new instantaneous $\Upsilon + \delta\Upsilon$ with existing average Υ, *OTS'* interface moves one step further in its evolution. Similar process of comparison and choosing the roadmap forms individual biography of *OTS*.

Note that from debated point of view, an information on the past of *OTS* contains the closed circle, which is being formed in interaction of two processes – deterministic move on y and random fluctuations on x, but not just one or a few individual loops ≈ $2lny$ that do not assume return to an original position. For such open trajectories development of *OTS* comes as clearly stochastic without chance to account accumulated data about the past.

Based on that it is reasonable to believe that quantitatively an information on the past depends on the number of circular integrals $\oint_C \delta\Upsilon$, where C is arbitrary circular contour of integration in the phase space of $\delta\Upsilon$. Going to the limit, in approximation of infinite number of closed circles (phase trajectories),

$$\Upsilon = \lim_{i\to\infty} \oint_{C_i} \delta\Upsilon$$

and is equivalent to the concept of phase volume that also can be calculated according to (1.12.a).

Note that this conclusion agrees with existing understanding that an integration transforms even originally Markovian process (with random change of x) to non-Markovian (regular changes in y) [7], at that memory (information on the past) is critical to choose direction of evolution.

Below, formally we will be using probabilistic definition of entropy as interface quantity knowing that all abovesaid is equally applicable for thermodynamic definition of entropy either.

So, in space M the greatest number of possible microstates W of given energy [8, 9] is calculated as $(d\Upsilon/dy)dy$, then by Gibbs' definition [8],

$$\Delta S = lnW = \ln|\ln y| \tag{1.14}$$

It is important to note that (1.14) coincides with a major part of differential entropy (1.23.a) discussed below.

From (1.14) follows that total entropy only growths in time

$$\frac{dS}{dt} = \frac{1}{\mathrm{y}\ln^2 y}(\ln \mathrm{y}\frac{dy}{dt}) \geq 0$$

irrelevant to the sign dy/dt, which confirms the validity of the approach taken through the nonnegativity character of entropy's dynamics in full correspondence with the 2nd law of thermodynamics [34].

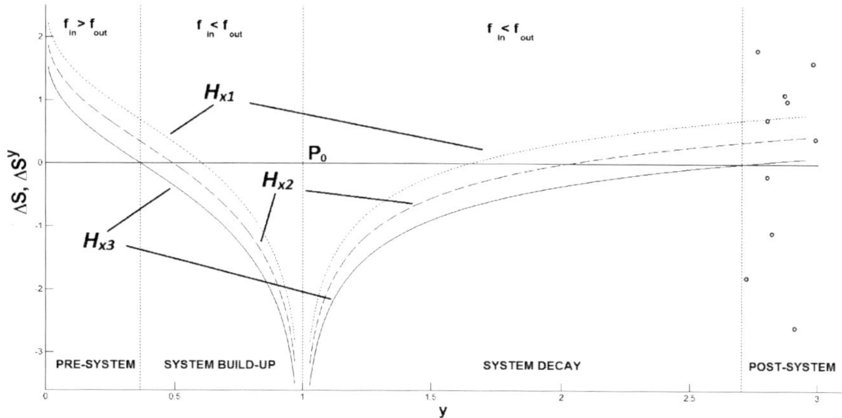

Figure 1.3. Differential ΔS^y and statistical ΔS entropy for an interface of *OTS* in dependence on rate of energy exchange y. In the plot, by an abscissa axis, dimensionless energy exchange rate $y = J/J_0$ is indicated, and by an ordinate axis, entropies ΔS^y и ΔS are indicated. Three curves in the range $H_{x1} > H_{x2} > H_{x3}$ are presented, where H_x is the probability distribution of entropy for random x, $H_x\,[0, \ln 2]$. The ratio between probability density for $f_{in} = f(y{=}y_{in})$ and $f_{out} = f(y = y_{out})$ is also shown, where $y_{in} = J_{in}/J_0$, $y_{out} = J_{out}/J_0$).

Also, from (1.14) stems that

$$\frac{d^2S}{dy^2} = -\frac{\ln y + 1}{y^2 \ln^2 y} \tag{1.15.a}$$

which at $y = GRP_L$ changes sign from plus to minus causing dS/dy to reach to local maximum

$$\left.\frac{dS}{dy}\right|_{y=\frac{1}{e}} = max \tag{1.15.b}$$

In addition, at $y = GRP_L$, from (1.14), $\Delta S = 0$.

It is easy to show that from (1.14) follows that at $y \to 0$, $y \to \infty$, quantity ΔS growths without limitation, which confirms previous conclusion that beyond above points operation of interface is not possible both at $y \leq 0$ (interface has not existed yet), and at $y \geq e$ (interface already has not existed), as is demonstrated in Figure 1.1.

Accounting abovesaid, the plot for ΔS looks as $H_x = H_{x3}$ in Figure 1.3 (discussed above).

1.1.6. Probability Density for Random *X*

Let $g(x)$ be the probability density function for random x meeting Kolmogorov's axiom, i.e., for $p(x) \geq 0$ at each $x \in X$, and $\int p(X)\,dx = 1$

$$g(x) > 0 \ for\ x \in [-1, 1] \tag{1.16}$$

$$g(x) = 0\ at\ x \notin [-1, 1]$$

then from (1.8.b) for each y at $y \neq SP$ accounting $|ln\ y|$,

$$f^y(x) = \frac{g(x)}{|\ln y|},\ x \in [-1,1] \tag{1.17a}$$

[10] and at $y = SP$

$$f^y(x) = \delta(x), \tag{1.17b}$$

where $\delta(x)$ is Dirac's delta function [11] and $|lny|$ is used to keep $f^y(x) \geq 0$, y is a parameter. Calculation of $f^y(x)$ is done in assumption that x is random at each fixed y. Then normalization condition нормализации for $f^y(x)$

$$\int_{-\ln y}^{\ln y} f^y(x)dx = 1 \tag{1.18}$$

Applying Leibnitz's rule for differentiation under integral [12], obtain that (1.18) is equivalent to

$$g(x) = \frac{1}{2} \tag{1.19}$$

Note that (1.18) at $|lny| < 1$ can be rewritten as

$$\int_{-1}^{1} f^{y}(x)dx = \int_{-\ln y}^{\ln y} f^{y}(x)dx + 2\int_{\ln y}^{1} f^{y}(x)dx \tag{1.20}$$

where is accounted that

$$|\int_{1}^{\ln y} f^{y}(x)dx| = |\int_{-\ln y}^{-1} f^{y}(x)dx|$$

and

$$|\int_{\ln y}^{1} f^{y}(x)dx| = |\int_{-\ln y}^{-1} f^{y}(x)dx| \cdot$$

An important detail of (1.20) is what the physical meaning of two terms in the right part is not the same. Ai $|lny| < 1$, integrals $|\int_{\ln y}^{1} f^{y}(x)dx|$ and $|\int_{-\ln y}^{-1} f^{y}(x)dx|$ deal with x-microstates, which are formally left unoccupied. Then, integral quantity in the range *[ln y, 1]* lacks equiphasic behavior with the quantity within the range *[0, lny]* and, therefore, cannot be described in the same way.

To support this semantic difference further, we will treat quantities for unoccupied microstates as imaginary ones. Based on that, rewrite (1.20) as

$$\int_{-1}^{1} f^{y}(x)dx = \int_{-\ln y}^{\ln y} f^{y}(x)dx + i \cdot 2\int_{\ln y}^{1} f^{y}(x)dx \tag{1.21}$$

The highlighted difference between ranges *[0, lny]* and *[ln y, 1]* will be additionally studied in chapter 1.3.

1.1.7. Differential Entropy of *OTS*

Although differential entropy S^y cannot be considered as the mathematically rigorous extension of statistical entropy S because of the known limitation $\lim_{\Delta \to 0} \ln(\Delta) \to -\infty$ [13], however, the difference ΔS^y obviously lacks this shortcoming, where Δ is the size of discretizing for probability density. Hence, further, we are only interested in the difference of differential entropy S^y between adjacent states. So, we will consider the dynamics of ΔS^y, taking as the second some arbitrary anchor value of S^y.

Consider an entropy probability distribution for an instantaneous random function $\delta \Upsilon$

$$\Delta S^y = -\int_{-1}^{1} f^y(x) \ln(f^y(x)) dx \tag{1.22}$$

Insert (1.16) to (1.22) and simplify, then at $y \neq SP$

$$\Delta S^y = H_x + \ln|\ln y|, \tag{1.23.a}$$

at $y = SP$

$$\Delta S^y \to -\infty \tag{1.23.b}$$

where an entropy of probabilistic distribution of x within the range *[-1, 1]*

$$H_x = -\int_{-1}^{1} g(x) \ln(g(x)) dx, \tag{1.24}$$

$H_x \in [0, \ln 2]$ while $\ln|\ln y|$ is major part of differential entropy matching (1.14).

From (1.23.a) follows that ΔS^y is the two-valued function with the roots

$$y_c^{\pm} = exp\left(\pm exp\left(-H_x\right)\right) \tag{1.25}$$

where "+" fits to the larger root and "-" the lesser one.

Note that (1.22) can be rewritten as

$$\Delta S^y = \Delta S^y{}_{out} + \Delta S^y{}_{in} \tag{1.26}$$

where independent

$$\Delta S^y{}_{out} = -\int_{-1}^{0} f^y{}_{out}(x)\ln(f^y{}_{out}(x))dx$$

$$\Delta S^y{}_{in} = -\int_{0}^{1} f^y{}_{in}(x)\ln(f^y{}_{in}(x))dx$$

$f_{in} = f(y{=}y_{in}),\ f_{out} = f(y = y_{out}),\ y_{in} = J_{in}/J_0,\ y_{out} = J_{out}/J_0).$

Then, (1.26) can be rewritten as

$$\Delta S^y = -\int_{-1}^{1} A^y(x)dx \tag{1.27}$$

where probabilistic anisotropy of interface

$$A^y(x) = \ln\left(\frac{f_{out}{}^{f_{out}}}{f_{in}{}^{-f_{in}}}\right), \tag{1.28}$$

Then, the roots $\Delta S^y = 0$ correspond to

$$f_{in} = W_L(f_{out}{}^{-f_{out}}) \tag{1.29}$$

with the plot in Figure 1.4.

Let $\Delta S^y \geq 0$, then from (1.27, 1.28) smooth (при $y \neq SP$) functions f_{in}, f_{out}, and accounting that $W_L(z)$ is growing function of z at $z \geq 0$, aggregate contribution of integrand $A^y(x)$ over the points where $A^y(x) \leq 0$ exceeds aggregate contribution over the points where $A^y(x) \geq 0$, and vice versa. The said above confirms to initialize $A^y(x)$ that at $\Delta S^y \geq 0$

$$f_{in} \geq W_L\left(f_{out}^{\;-f_{out}}\right) \tag{1.30.a}$$

while at $\Delta S^y \leq 0$

$$f_{in} \leq W_L\left(f_{out}^{\;-f_{out}}\right) \tag{1.30.b}$$

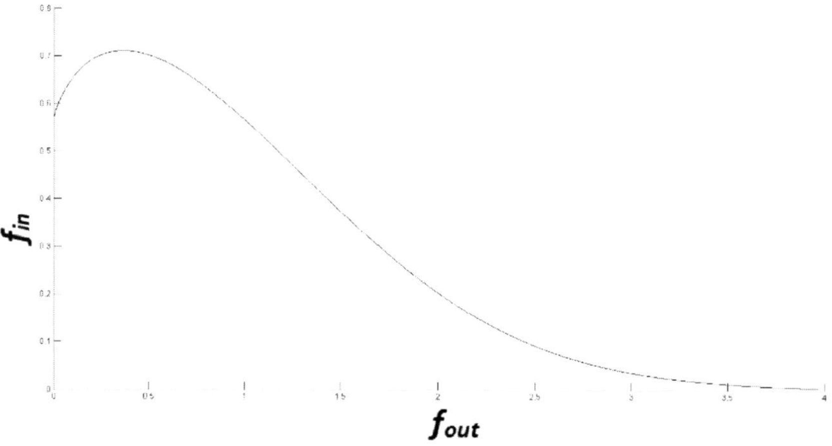

Figure 1.4. Dependence of probability density inwards energy flow f_{in} on the outwards energy flow f_{out} at $S^y = 0$ for interface OTS, $f_{in} = W_L(f_{out}^{-f_{out}})$. In the plot, by the abscissa axis density f_{out} is indicated, by the ordinate axis density f_{in} is indicated. Here, $f_{in} = f(y=y_{in})$, $f_{out} = f(y = y_{out})$, $y_{in} = J_{in}/J_0$, $y_{out} = J_{out}/J_0$.

Note that (1.27) does not have any limitations in terms of the irreversibility or the time scale of the thermodynamic process.

1.1.8. Zonal Structure of Entropy

As it is seen from (1.23a), interface entropy includes the term Hx, which is an entropy probability distribution for the compact support $x \in [-1, 1]$

$$H_x = -\int_{-1}^{1} g(x)\ln(g(x))\,dx \tag{1.31}$$

Entropy H_x is non-negative quantity in the range

$$H_{x\ min} \le H_x \le H_{x\ max}, \tag{1.32}$$

where borders $H_{x\ min}$, $H_{x\ max}$, generally, depend on logarithm base z [14].

As H_x is bound, then according to (1.23.a) it further leads to the limited range for

$$H_{x\ min} \le S - ln|lny| \le H_{x\ \max} \tag{1.33}$$

or, rewriting (1.25),

$$y_c^{\pm} = exp\left(\pm exp\left(S - H_x\right)\right) \tag{1.34}$$

Observe from (1.34) that S is the two-valued function, where the lefthand branch is $y = exp\left(-exp\left(S - H_x\right)\right)$ and the right-hand one is $y = exp\left(exp\left(S - H_x\right)\right)$.

Then, the full set of admissible values y–S consists of the two ranges. To be exact, the left-hand range is $[y_L^-, y_L^+]$ and the right-had one is $[y_R^-, y_R^+]$, where $y_L^- = exp(-exp(S - H_{xmin}))$, $y_L^+ = exp(-exp(S - H_{xmax}))$, $y_R^- = exp(exp(S - H_{xmax}))$, $y_R^+ = exp(exp(S - H_{xmin}))$.

In considering model, minimum dissipation fits $H_x = 0$. Then, the pair of the outmost entropy curves is $y = exp(\pm exp(S))$, which at $S = 0$ yields the roots of entropy S equal to $y_L^- = GRP_L$ and $y_R^+ = GRP_R$.

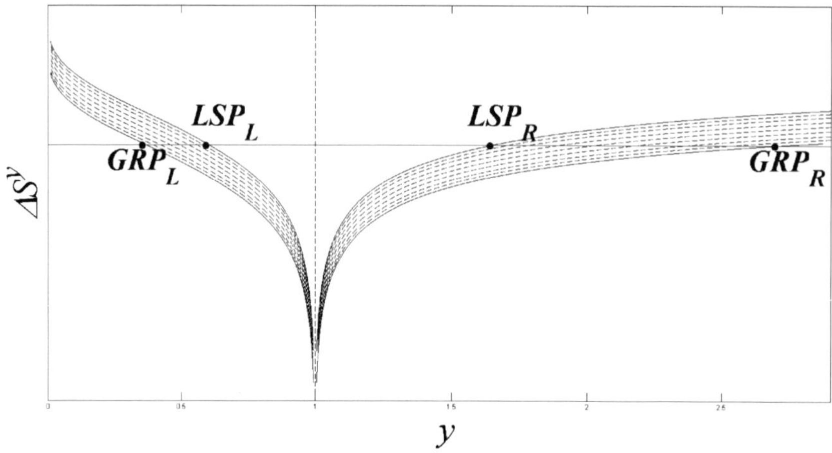

Figure 1.5. Dependence of *OTS* differential entropy $\Delta S^y(H_x,y)$ on rate of energy exchange *y*. In the plot, by the abscissa axis an energy exchange rate $y = J/J_0$ is indicated, by the ordinate axis change of entropy ΔS^y is indicated. Plot consists of two pairs of curves, one pair fits the pair of the left roots of ΔS^y (GRP_L and LSP_L), other pair fits the right roots (GRP_R и LSP_R). Allowable ΔS^y is found within two dashed areas. The roots of entropy ΔS^y match the spectrum nodes for the fundamental (GRP_L, GRP_R) and the second (LSP_L, LSP_R) harmonics of *OTS* discrete spectrum, considered in chapter 1.2, 2.2.

To calculate the innermost border curves, we should know value of $H_{x\,max}$, which physically corresponds to the state of the highest disorder possible when a random variate *x* has uniform distribution [14]. In the given model it appeals to the situation, at which we have a minimum of knowledge on state of *OTS* a priori. According to (1.19) appropriate probabilistic density $g(x) = ½$, then $H_{x\,max} = ln\ 2$, roots $y_L^+ = LSP_L^{1/2}$ and $y_R^- = LSP_L^{-1/2}$, where meaning of *LSP* will be highlighting in chapter 2.2.

Now, based on the found entropy roots, we may write

$$\lim_{H_x \to \min} S_0 = \exp(\pm 1) \tag{1.35.a}$$

$$\lim_{H_x \to \max} S_0 = \exp(\pm \frac{1}{2}) \tag{1.35.b}$$

where S_0 denotes the set of the roots of entropy *S*.

So, the feasible solutions for entropy as follows from (1.33) are localized inside the area bound by the curves

$$H_{x\,min} + ln|lny| \le S \le H_{x\,\max} + ln|lny| \tag{1.36}$$

as shown in Figure 1.5.

It is important to note that $y-$roots of entropy are equal to the first and second harmonic of discrete spectrum for evolving *OTS* (*GRP* and *LSP*), considered in the chapter 1.2.

1.1.9. Discussion for Chapter 1.1

In this chapter, we substantiated the suggested method for analysis dynamics of *OTS* with an unlimited number of conserved links, presented suitable mathematical formalism, converted formalism into a system of infinite number of *CE* (1.2) named *RECE*, found an analytic solution of the system (1.2), described a phase space, and provided detail profile of *OTS* entropy .

Found solution (1.12.a) can be considered as analog of *CE* (1.1) in coordinates $y - \Upsilon$. In its meaning, solution Υ is an integral efficiency of energy exchange through *OTS'* interface.

Properties of function Υ allow to describe a full lifetime cycle of *OTS* on the high level of abstraction. It is worth noting that the sufficiently universal essence of the model and the inbuilt *RECE* formalism provide arguments in favor of the capacity of the model to adequately describe a non-equilibrium energy exchange in an open system.

It should be noted that function Υ, by its definition, has any meaning only for open systems.

Firstly, from (1.7), it follows that $\Upsilon \sim dU/Q$, which directly points out at Υ as an interface between transferred quantity Q and change of system quantity dU, such a kind of interface is only possible at condition of system's openness. Secondly, existence of Υ is principally impossible without energy flow as Υ is based on the flow exchange $OTS \leftrightarrow$ externals, which makes any sense for open system only.

From a formal standpoint, evolution reduces to competition between flow y and random impact of interface factor x. If $dy = 0$ (fixed y), the influence of x comes to the forefront and totally controls the course of evolution.

In the stationary regime (при $y = 1$), the range of x formally is being shrunk to the only point. Clearly, it is one of the reasons for the stability of the stationary regime, and it explains why it is a "genuine stationary" regime in contradiction to quasi-stationary regimes occurring at $y \neq 1$, when even at zero flow change $dy = 0$, value x, nonetheless, is taken randomly from the finite x-range.

It is worth noting that in discussing theory, both paths towards the greater y and the lesser y are always (excepting stationary point *SP*) separated. The reason for this lies in what to switch from one branch to another, instant efficiency $\delta\Upsilon$ mandatorily must go through the stochastic zone ($dy = 0$). So, the probability of returning to the same state at the same y is vanishingly small. As to return of *OTS* to previous state from the stages "Non-System" или "Post-System" (Figure 1.1), then such return is irreversible by definition, as *OTS* firstly must go through the total disintegration and can return to system trajectory ($0 \leq y \leq GRP_R$) only as brand new *OTS*.

Considering evolution of *OTS* is based on existence of probabilistic anisotropy of interface $A^y(x) \neq 0$ (1.36) in the space *M*. Pay reader attention to proximity of mathematical form of (1.36) to results [41], where factor $ln\ [\pi(II \rightarrow I)/\pi(I \rightarrow II)]$ establishes macroscopic agreement with 2nd law of thermodynamics, while quantities $\pi(II \rightarrow I)$ and $\pi(I \rightarrow II)$ are defined as forward and reverse transition probabilities.

Functions efficiency Υ and entropy S are bound to each other by (1.21). Besides, connection between Υ and differential entropy already exists that is identical to (1.21) up to an arbitrary constant.

Then, entropy S is logically to be considered as an indicator of uncertainty of existing knowledge on the state in which the *OTS* interface will be in the next moment of time, while Υ describes that volume of data from which this choice will be made. Hence, S accounts local in time change in state of interface, but Υ does it for integral change of state, accumulating full history of interface's evolution.

On the other hand, the greater part of energy is being dissipated and converted to a low-structured state (fits to the greater S) outside *OTS*, the less efficiency Υ in the exchange process *OTS* ↔ externals will be, which confirms the opposite essence in changes S and Υ.

Conclusion of Chapter 1.1

1. Infinite system (1.2) is a mathematically correct model of *OTS* with an unlimited number of conserved links.
2. System (1.2) presumes transformation to equivalent stochastic differential equation *RECE* (1.6.a).
3. The additive structure of Υ (dependence on y and independence on sign of dy) supports accumulation of memory on changes in state of *OTS*, making energy development of *OTS* the non-Markovian process. It allows distinguishing the new instant $\Upsilon + \delta\Upsilon$ from existing average Υ and thereby creating a basis for the primitive cognition function of *OTS*. A similar process of comparison and choosing the roadmap forms the individual biography of *OTS*.
4. Solution (1.12.a) is bound with entropy of *OTS*; it is confirmed for the statistical (1.14) and differential (1.23.a), as well as thermodynamic (3.26) definitions of entropy.

Chapter 1.2

The Spectrum of Energy Exchange in *OTS*

1.2.1. Bound Value Problem for Υ

In this chapter, the spectral response to evolutionary dynamics of *OTS* within the model presented above will be studied. With this purpose, the boundary value problem will be solved for the operator of *OTS'* interface obtained earlier (1.12.a). In the light of the intercoupling of Υ and entropy *S* established previously, the *y*-spectrum of entropy and possible scenarios of *OTS'* evolution stemming from here will also be analyzed.

Pay attention that Υ by construction does not have any meaning beyond the range *[0, GRP$_R$]* as it becomes negative. So, solution $\Upsilon(y)$ is localized on a finite interval, and we come to a two-point boundary problem on interval *[0, GRP$_R$]* with solution as per (1.12.f).

$$\Upsilon(0) = \Upsilon(e) = 0 \tag{2.1}$$

At that, in chapter 1.1, the continuous variation of quantity k_X was noted (1.12.c).

On other hand, equating (1.12.a) к нулю, find roots of *n*-order

$$1 = \pm n \ln y_n \tag{2.2}$$

or

$$y_n = \exp[\pm \frac{1}{n}] \tag{2.3}$$

with suitable harmonics

$$\Upsilon_n = y_n(1 - \ln y_n) \tag{2.4}$$

where $n = 1,2\ldots$.

The same result can be obtained if to use использовать Lagrangian

$$L = T - U = QE_A - U = 0$$

where $T = QE_A$ is kinetic energy, U is potential energy. Using standard condition

$$\frac{dL}{dt} = \frac{d}{dt}\frac{dL}{dy} = 0$$

we have $\Upsilon = 0$ again and n-roots as shown in (2.3). It confirms that (2.3) deals with the points of equilibrium (*PoE*) for the process of energy exchange in *OTS*. So, based on the abovesaid, factor

$$k = \frac{\Upsilon_n}{y_n} = k_n = 1 \pm ln\, y_n = 1 \pm \frac{1}{n} \tag{2.5}$$

is a discrete quantity.

So, there exists dualism in the description of factor k, revealing in the appearance of the continuous formalism (1.12.c) and the discrete one (2.5). The essence of such dualism will be discussed in Chapter 2.3 , in which the role of time asymmetry as a factor of *OTS* evolution is shown.

Relation (2.5) corresponds to $|ln\ y| \leq 1$, where the latter determines y within the range $[GRP_L, GRP_R]$.

The discrete spectrum for Υ is shown in Figure 2.1. As demonstrated in chapter 2.2, the effects observed in the first three spectral nodes (2.5) have transparent physical meaning. The physical effects in higher order nodes are not significant compared to the first three and look similar.

Combined plots for instantaneous efficiency of energy exchange $\delta\Upsilon$ (1.8.b), integral efficiency Υ(1.12.a), and total exchange energy T (1.12.d) are given in Figure 2.2.

Notice that, formally, (1.12.d) allows to define an independent discrete y-spectrum for total energy exchange T. However, additional research confirmed that this spectrum does not have such crucial evolutionary context as the spectrum of Υ(1.12.a). It will also be confirmed by the results of chapter 2.1. In summary, T-discretization is not considered in further text.

1.2.2. Quasi-Continuous Spectrum of Entropy

As was discussed in paragraph 1.1.8, quantity H_x in (1.31) is bounded. Its maximum is realized for unified probability distribution, in this case density of random x is $g(x) = ½$, and $H_{x\,max} = ln\ 2$. Typically, in minimum $H_{x\,min} = 0$, which corresponds to probabilistic state of largest possible physical order [15]. So, (1.32) can be rewritten as

$$0 \le H_x \le \ln 2 \tag{2.6}$$

Plug in (2.3) in (1.23.a), then

$$\Delta S_n = H_x - \ln n \tag{2.7}$$

Dependence (2.7) means that solution exists within two domains shown in Figure 1.5, i.e., y-dependence of entropy takes the quasi-continuous form.

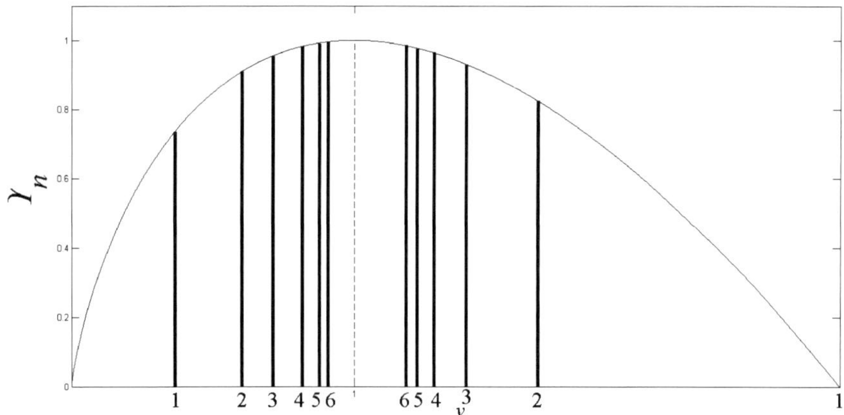

Figure 2.1. Discreteness of spectrum for integral efficiency of energy exchange Υ. In the plot, by an abscissa axis, a unitless energy exchange rate $y = J/J0$ is indicated, and by an ordinate axis, the integral efficiency of energy exchange Υ. The first six pairs of harmonics Υn are schematically shown by thick vertical segments in the range $GRP_L \le y \le GRP_R$. Conjugated (paired) harmonics are positioned asymmetrically relatively *SP* (shown in dash).

Discrete solution (2.7) is

$$\begin{cases} n = 1, \ y_1^{\pm}, \ H_x = H_{x\,min} \\ n = 2, \ y_2^{\pm}, \ H_x = H_{x\,max} \end{cases} \tag{2.8}$$

or in designations of paragraph 1.1.8.

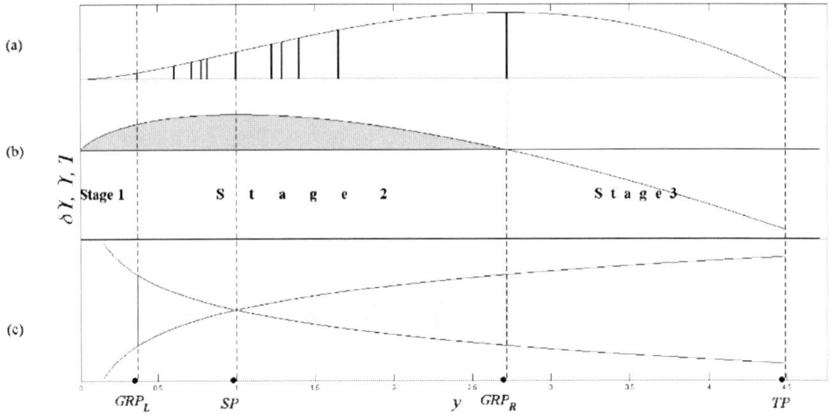

Figure 2.2. Total exchange energy T (a), average efficiency of energy exchange Υ (b), instant efficiency of energy exchange $\delta\Upsilon$ (c) in the phase space for possible microstates of $\delta\Upsilon$ are shown. In the plot, by an abscissa axis, a dimensionless energy exchange rate $y = J/J_0$ is indicated. The continuous spectrum is in Stage 1 ($0 \leq y \leq GRP_L$) and Stage 3 ($GRP_R \leq y \leq TP$). The discrete spectrum is in Stage 2 between the points GRP_L and GRP_R (shown in the light grey in the panel (c)), the range with positive Υ is between the points 0 and GRP_R (shown in the dark grey in the panel (b)). The discrete energy levels T_n are shown by the thick vertical segments in the upper panel (a). The point TP marks termination of the whole cycle for OTS energy development.

$$\begin{cases} GRP_L \leq y \leq LSP_L \\ LSP_R \leq y \leq GRP_R \end{cases} \tag{2.9}$$

that is why the y-borders in (2.6) are the first and second harmonics of the discrete y-spectrum (2.3).

By logic of [16], maximum H_x corresponds to the largest number of random closed cycles. Appropriately, minimum H_x deals with the lesser of those. Besides, if the process demonstrates maximum production of entropy, then this process corresponds to the most probable and fast path of evolution.

In this theory, maximum production of entropy meets maximum rate of dS/dy, which occurs at $y = GRP_L$ (1.15.b). From (2.7) $H_x (y = GRP_L) = 0$. In Figure 1.3, it accommodates the curve ΔS^y, which intersects the S axis in the points matching the fundamental harmonics $y_1^- = GRP_L$ and $y_1^+ = GRP_R$.

So, the most probable and fast path of *OTS* evolution is realized for the fundamental harmonic of the discrete spectrum and matches curve $\Delta S^y(y)$ at $H_x = H_{x3}$ in Figure 1.3.

For point $y = e^{-1/2} = LSP_L$ from (2.7), it follows that $H_x (y = LSP_L) = H_{x\ max} = ln2$. For LSP_R, $y = e^{1/2}$. It qualifies LSP_L (LSP_R) as the maximum possible point from an entropy standpoint (further will be shown that actually y can exceed LSP_L or LSP_R). The physical meaning of this limitation will be discussed below. So, keeping $H_x = H_{x3}$ guarantees the least probable and slowest path of evolution. In Figure 1.3, it is highlighted by the curve ΔS^y that crosses the y-axis in the points matching the second harmonic $y_2^- = LSP_L$ and $y_2^+ = LSP_R$.

So, evolutionary scenarios meeting (2.7) lie within two intervals, where the y-borders are fundamental and second harmonics. At that, continuous solutions at $1 < n < 2$ occupy a niche between y_1 and y_2 with a diminishing probability of realization if to move from y_1 to y_2. If y violates conditions (2.7, 2.8), then in the first approximation, evolutionary scenarios do not exist. However, thorough scrutiny in chapter 2.2 removes this limitation.

We should notice that, within considering model realization of evolutionary scenarios, which lack phase $\Delta S_n > 0$, is not possible. It follows from (1.31), as in this case the probability for realization of random x is less than ½, which contradicts the assumption in chapter 1.1 about uniform distribution of x in the range $[-1, 1]$. From a physical viewpoint, the absence of phase $\Delta S_n > 0$ can testify that ordering in *OTS* cannot start if the previous phase of chaotic energy accumulation ($\Delta E_n > 0$) for some reasons was not involved.

Also, it should be noted that the employed condition $H_{x\ min} = 0$ equivalent to the state of highest possible physical order is to be understood not in the meaning of predetermination for realization of one or another microstate of *OTS* in the point $y = GRP = e^{\pm 1}$. This condition indicates that in order to go through *GRP*, the system must alternate the form of spectrum (from continuous to discrete or vice versa), and any other variants of evolution are not possible (singularity of *GRP* was highlighted in [17]). Through these lenses, maximum H_x in the point $y = LSP = e^{\pm 1/2}$ verifies that from probabilistic viewpoint, passing by node *LSP* is impossible, which is in line with the mentioned absence of evolutionary scenarios at $n > 2$. However, if fact, as shown in Chapter 2.2, it only means that probability to pass by *LSP* is below 50%, however implementation of evolutionary scenarios is still possible.

Consequently, in this paragraph, we detected an interrelation between possible scenarios of *OTS* evolution and properties of the spectral nodes.

1.2.3. Mechanism of Υ Discreteness

As follows from (2.3), constraints imposed on Υ in range $[0, GRP_R]$, factually assume quantization of unit Υ-interval $I = [0, 1]$. Let's analyze the meaning of this conclusion.

For clarity, the focus below will be on a segment in the upper part (between $-ln\,y$ and y-axis) of phase space $x - y$ marked by a rectangle in Figure 2.3 (a). The scaled-up version of that rectangle is shown in Figure 2.3 (b). All considerations presented below are also equally applicable for the lower part of phase space $x - y$ (between the curve $ln\,y$ and axis y) with an appropriate change of the sign in $ln\,y$ and 1.

At $y < GRP_L$, holds $|\delta\Upsilon| > 1$ (Figure 2.3 (a) dark-grey domain), as a result, in this domain current $\delta\Upsilon$, to one degree or another, can be out of the interval I, and no limitations on parameters of the evolutionary process are imposed.

However, at $y > GRP_L$ quantity $|\delta\Upsilon| < 1$ (Figure 2.3 (a), light-grey domain) and the situation would be quite different. Now, all $\delta\Upsilon$ are within I. It leads to splitting I into two parts, I_r and I_r. Consider the meaning of these parts. From (1.8.b) follows that condition $|\delta\Upsilon| = \int dU/Q = 1$ (horizontal dash line in Figure 2.3 (a)) means that change in incoming energy flow Q matches change in internal energy dU caused by Q. In the remaining cases, i.e., at $|\delta\Upsilon| \neq 1$, only the portion of flux Q is utilized in the changes of energy dU while the residual energy is the unstructured thermal one that is taken off *OTS*.

So, the range I_r addresses the portion of energy exchange process microstates utilized by $OTS \sim ln\,y$ in the form of internal energy dU, whereas the range I_i deals with the residual microstates in the volume $\sim 1 - |ln\,y|$ to support droppable thermal energy (Figure 2.3 (b)).

At that, I_r supports changes $\delta\Upsilon$ in the range $[0, ln\,y|]$, while I_i formally does the same in the range $[1 - |ln\,y|, |ln\,y|]$, where $I_r + I_i = I$ (Figure 2.3). The first portion is the actual value that forms changes in dU. The second portion deals with energy values that were used before but have become useless beyond some boundary *y*.

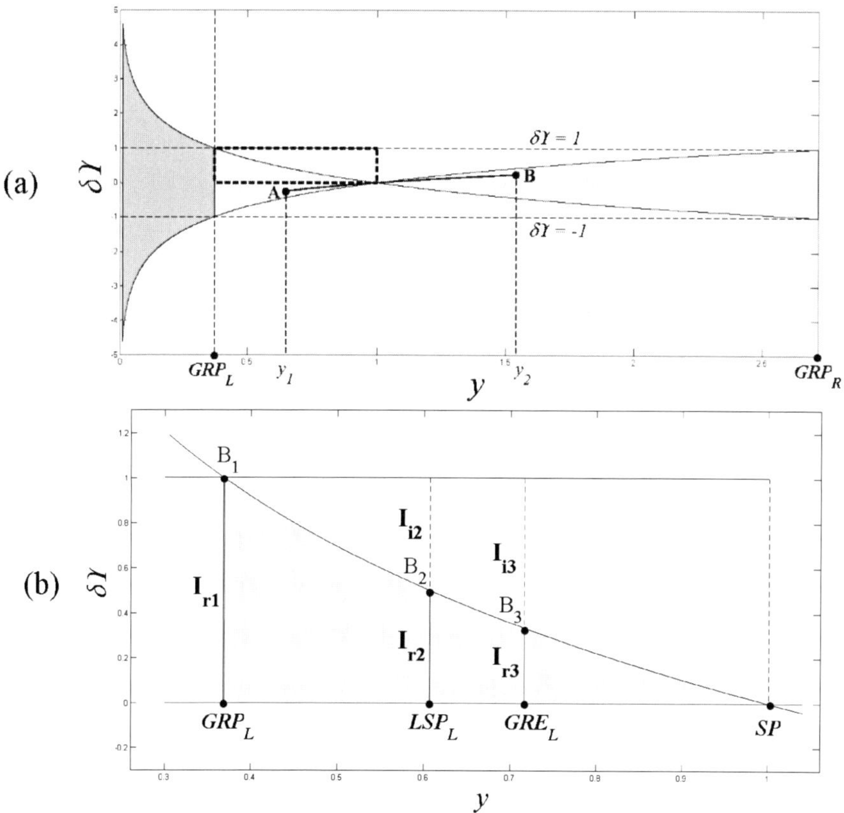

Figure 2.3. (a, b). Phase space (a) and confinement area for $\delta\Upsilon$ (b). In the plot, by an abscissa axis an energy exchange rate y is indicated, and by an ordinate axis the instant efficiency of energy exchange $\delta\Upsilon$ is indicated. The panel (b) is the scaled-up version of the rectangle marked by the thick dashed lines in panel (a). Area $|lny| > 1$ at $y < GRP_L$ is shown in the dark grey in the panel (a). The trapping area with $|lny| \leq 1$ at $GRP_L \leq y \leq GRP_R$ is shown in the light grey in panel (a). The support ranges for I_r ($B_1GRP_L = I_{r1}$, $B_2LSP_L = I_{r2}$, $B_3GRE_L = I_{r3}$) are in the grey area while the support ranges for I_i (shown in dashed lines) are in the white area in panel (b). The essence of I_r, I_i as well as the points *GRP*, *LSP*, and *GRE*, is explained in greater detail in the chapter 2.2, and below. All shown quantities are dimensionless.

In the foregoing sense, there exists a principal qualitative difference between I_r and I_i, which play opposite roles in the evolution performance of

OTS. In order to emphasize this fact further, we will assume that function $1 - \ln y$ is imaginary with changes along the axis orthogonal to y [18].

Discussion on physical meaning of функции $1 - \ln y$ will be conducted in chapter 2.1.

Accounting above, normalizing condition for I_r and I_i comes as

$$\ln^2 y + (1 - |\ln y|)^2 = 1 \tag{2.10.a}$$

$$2\ln y(|\ln y| - 1) = 0 \tag{2.10.b}$$

from which follows (2.3).

So, the physical mechanism of Υ confinement is based on the quantizing of reference unit range I and formally stems from the normalizing condition for this range (2.10).

1.2.4. Mechanism for Quasi-Continuity of Entropy

All possible entropy ΔS^y (1.23.a) are defined by the diapason H_x for random x (1.31), so localization (quasi-continuity) of ΔS^y from this point of view is easily understandable. Let's analyze the occurrence of a quasi-continuous spectrum from probabilistic stands.

For the discussed model with uniform x-distribution, the probability of falling random x into some interval is proportional to the length of this interval (further *UPD* assumption). Therefore, the probability p_{ev} to fall into the interval $I_r = [0, \ln y]$ (grey area in Figure 2.4 (b)) is proportional to the length of this interval $|\ln y|$, whereas the probability p_{nev} to fall into the complementary interval I_i (white area in Figure 2.3 (b)) is defined by $1 - |\ln y|$. Here and below, p_{ev} deals with the probability of an evolutionary scenario, while p_{nev} deals with the probability of a non-evolutionary scenario in space y - $\delta\Upsilon$ (in terms of [16]).

Hence, ratio

$$\vartheta(n) = \frac{p_{ev}}{p_{nev}} = \frac{|\ln y|}{1 - |\ln y|} = \frac{1}{n-1}, \quad n \neq 1 \tag{2.11}$$

serves as an indicator for the relative value of p_{ev} and p_{nev}.

To compare physical conditions in the discovered phases of energy evolution *Pi*, look at Table 2.1 below. This table was created based on (2.11) and a simple calculation of the sign for $v = dY/dy$.

We see that the combination of physical conditions determined by the factors v, ϑ, ζ is unique in each phase *Pi*.

At $n < 2$, holds $\vartheta > 1$ and $p_{ev} > p_{nev}$, while at $n > 2$, holds $\vartheta < 1$ and $p_{ev} < p_{nev}$. If $n = 2$ $(y = y_2)$, then $p_{nev} = p_{ev}$, therefore, in this case, the chances for evolutionary and non-evolutionary development of *OTS* are the same. So, at $y > y_2$, the probability for realization of the non-evolutionary scenario looks more preferable compared to the range $y < y_2$, where the evolutionary development predominates.

Table 2.1. Unique physical conditions in the phases of energy evolution. The sign of v, as well as the value of ratios ϑ and ζ, is shown

Physical essence of phases			
Phase #	v	ϑ	ζ
Phase 1	*Positive*	$1 < \vartheta < \infty$	$p_{ev} > p_{nev}$
Phase 2		$0.5 < \vartheta < 1$	$p_{nev} > p_{ev}$
Phase 3		$0 < \vartheta < 0.5$	$p_{ev} = 0$
Phase 4	*Negative*		
Phase 5		$0.5 < \vartheta < 1$	$p_{nev} > p_{ev}$
Phase 6		$1 < \vartheta < \infty$	$p_{ev} > p_{nev}$

Perform direct calculations in (2.11). At $|ln\ y| > |ln\ LSP|$, $p_{ev} > p_{nev}$ and, as a result, quantity $\vartheta > 1$, whereas at $|ln\ y| < |ln\ LSP|$, $p_{ev} \leq p_{nev}$ and $\vartheta \leq 1$. Hence, at $|ln\ y| > |ln\ LSP|$, in a probabilistic sense, the evolutionary scenario has the higher chances for realization. If $y = LSP$, then the chances for realization of evolutionary and non-evolutionary scenarios are the same. Finally, at $|ln\ y| < |ln\ LSP|$, most probably, *OTS* takes the path of non-evolutionary development. Therefore, intervals $[GRP,\ LSP]_L$ and $[LSP,\ GRP]_R$ include the majority of possible evolutionary scenarios.

In *LSP*, (2.11) yields $\vartheta\ (n = 2) = 1$, which provides static balance between probabilities of "success" and "failure" that corresponds to maximum variance $Var\ (x)$ (Figure 2.4).

As a result, evolution follows the way of the most probable scenario, thereby imposing limitations on entropy ΔS. Downsizing and splitting phase volume of energy exchange between utilized and non-utilized microstates appears in the formation of a quasi-continuous spectrum of entropy ΔS. In this sense, Υ-discretization and S-quasi-continuity are bound. Note that (2.7), formally, prevents the occurrence of evolutionary scenarios, in which entropy lies out of dashed domains in Figure 1.5.

Also, highlight that at $|lny| > 1$ relation (2.11) altogether with n, obviously, does not have any sense. It confirms that at $|lny| > 1$ $(y \leq GRP_L)$, the existence of a discrete spectrum is impossible.

1.2.5. Eigenvalues for Operator of *OTS* Evolution

Note that (1.12.a) takes rate y and converts it into efficiency Υ. In this sense, (1.12.a) can be considered as an operator equation for y.

So, write an eigenvalue problem as

$$L(y) \pm r\,y = 0 \tag{2.12}$$

where $L(y)$ is a nonlinear operator and eigenvalue r is some, generally, complex number.

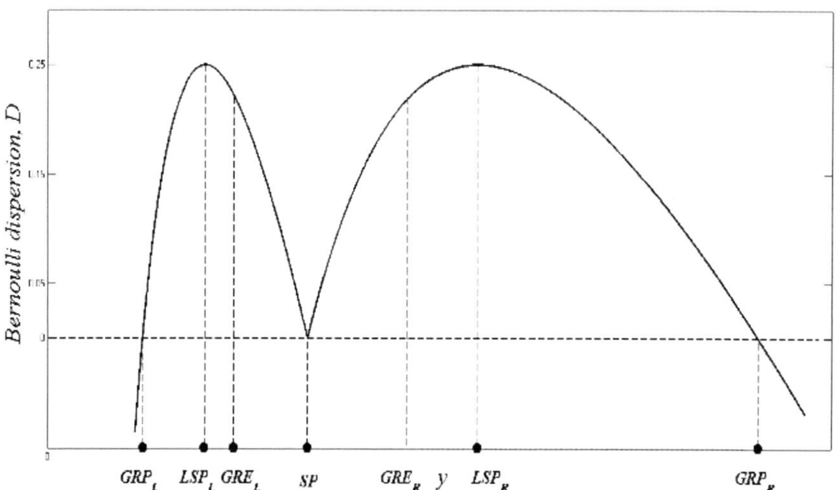

Figure 2.4. Dependence of Bernoulli's variance D on rate y. In the plot, by an abscissa axis an energy exchange rate y is indicated, and by an ordinate axis Bernoulli's variance is indicated.

In (2.12), replace *L(y)* with (1.12.a), so we obtain

$$y - y\ln y \pm r\,y = 0$$

and r matches k_X, introduced in (1.12.b), i.e.,

$$r = k_X = 1 - sign\,(\ln y)|\ln y| \tag{2.13}$$

As above, we limited interval y for discrete solution by $GRP_L \le y \le GRP_R$, and then from (2.13) it follows that the real part of r is also limited

$$0 < \mathrm{Re}(r_n) \le 2 \tag{2.14}$$

while the imaginary part can be of any value.

In matrix form, solution (1.12.a) for range of discreteness can be written as

$$\Upsilon_n = R_n \times Y_n \tag{2.15}$$

where R_n is vector-column of r_n, and Y_n is vector-row of y_n.

Thus, we come to the conclusion that the real eigenvalues r_n of operator *L(y)* are bound. The real part of r_n forms an infinite counted set (that is why r_n is better defined as an eigenvector of operator *L(y)*) within interval *(0, 2]*, with appropriate y-interval $[GRP_L, GRP_R)$. At that, the imaginary part of r_n does not have limitations.

The above said means that in y-interval $[GRP_L, GRP_R]$, the resultant action of operator L on flow y is reduced to multiplying by imaginary r_n, which is equivalent to a change in value and direction of y. In terms of k-basis, such transformation is equivalent to the change in form and size of a triangle at its rotation (Figure A.4).

Further it will be shown that eigenvector r_n plays a critical role in considering model as it ties physical and geometrical parameters of solution.

1.2.6. Pairing Property of Discrete Spectrum

In this theory, $\delta\Upsilon$ is the non-gradient function in the x-y phase space due to the existence of form (1.8.b) and, generally, $\oint_C \delta\Upsilon(x, y) \neq 0$ for any $y \in [0, GRP_R]$. Obviously, by making x fixed x_f, we would obtain that for all allowable $\{x_f, y\}$-points, $\oint_C \delta\Upsilon(x_f, y) = 0$, $\delta\Upsilon$ is a gradient field and Υ is independent of the path, i.e., Υ is a function of state.

However, even if we do not have the closed contour, i.e., in the case $x = x_f$ and $y_1 \neq y_2$, there are still the points for which holds

$$\oint_C \delta\Upsilon(x_f, y_i) = 0 \tag{2.16}$$

Condition (2.16) can occur for the $\delta\Upsilon$-contours that cross the points with the logarithmic symmetry in respect of line $y = 1$, or, in other words, the points where the natural slope of the curve $\pm ln\ y$ is cancelled by the right choice of y_1 and y_2. Mathematically, it takes place for any pair of y-points meeting condition

$$y_1 \cdot y_2 = 1 \tag{2.17}$$

Then, for any $x_f \in I$ and the starting point y_1, closed contour C degrades to the curve $x_f \cdot ln\ y$ crossing the points y_1 and y_2 that satisfy (2.17).

Consequently, (2.17) converts to

$$\int_{y_1}^{y_2} \delta\Upsilon(x_f, y)dy = 0 \tag{2.18}$$

along the curve AB shown in Figure 2.3 (a).

Obviously, if bend is zero and horizontal segments of $\delta\Upsilon$ are parallel to axis y, corresponding y-points would have usual axial symmetry relative to $y = 0$, which is not considered here.

Since any point in the range $[GRP_L, SP]$ has its logarithmic inverse (2.17) in the range $[SP, GRP_R]$, then these ranges have the perfect logarithmic symmetry relative to the line $y = 1$, and the points of discrete spectrum can come only in pairs.

1.2.7. Discreteness-Optimality Intercoupling

In agreement with (1.8.b), compose a general functional for *OTS* evolution as

$$\int_0^e F(x, y, \delta\Upsilon)\, dy = \int_0^e (\delta\Upsilon - x \ln y)\, dy \tag{2.19}$$

Based on what was said in the previous sections, it is reasonable to believe the close value for the quantities $\delta\Upsilon$ and x is within the discreteness interval $[GRP_L, GRP_R]$. So, it is logical to seek the unknown function f from the form

$$\Phi = \int_0^e (f - f \ln y) dy$$

Then, equating

$$\frac{\partial\Phi}{\partial f} = 1 - \ln y \tag{2.20}$$

to zero, we come to the discreteness condition (2.3) that at the same time minimizes (2.19) [19].

So, the appearance of the discreteness in this model is the natural way to optimize energy exchange in *OTS*.

Conclusion of Chapter 1.2

1. The $\Upsilon(y)$-spectrum of *OTS* simulated by equation *RECE* comes in two forms: continuous and discrete. A continuous spectrum is observed in the low ($y < GRP_L$) and high ($y > GRP_R$) ends of y-spectrum.
2. Spectrum of entropy S gets quasi-continuous character, occurring between S-curves crossing axis y in the points corresponding to the fundamental and second harmonic of Υ-spectrum. The more probable

scenarios of evolution are attracted towards the fundamental harmonic, while the less probable towards the second harmonic. Outside the interval $[y_1, y_2]$, the probability of an evolutionary scenario gives way to that of a non-evolutionary scenario.

3. The eigenvector of interface operator r_n takes an infinite number of values in the real range *(0, 2]*, while imaginary part of r_n can have any value.
4. Evolution follows the way of the most probable scenario of energy development of *OTS*, which imposes limitations on both the ranges for existence of entropy *S* and efficiency *Ƴ*. The most probable scenario ultimately is determined by the split of the phase volume, manifesting by the ratio of exchange energy between the real *ln y* and imaginary *1 – lny* part.
5. The reason for the *Ƴ*-discreteness is quantization of a unity range *[0, 1]*, which emerges at applying the normalization of an orthogonality between real *ln y* and imaginary *1 – lny* part.
6. The reason for the emergence of quasi-continuance in the entropy spectrum is the confinement of a distribution for the random *x* in the range *[-1, 1]*.
7. Any point in the range $[GRP_L, SP]$ has its logarithmic inverse (2.17) in the range $[SP, GRP_R]$, and the points of the discrete spectrum can come only in pairs.
8. The appearance of the discreteness in the spectrum of energy evolution is the result of optimizing the structure of energy exchange in *OTS*.

Chapter 1.3

The Connection of *RECE* to the Relations of Nonequilibrium Thermodynamics

1.3.1. Substantiation

In this chapter, the system analysis of *OTS* with an infinite number of conserved links is continued. The emphasis is on the investigation of the uncertainty in the energy state of *OTS*.

It is known that in classic physics there exists a general thermodynamic uncertainty relation (*TUR*) between local temperature ΔT and accumulated energy ΔE that restricts their simultaneous measurement. This relation is analogous in its meaning to the relation in quantum physics for simultaneous measurement of position and momentum [20].

TUR can be expressed as

$$\Delta E \cdot \Delta(\frac{1}{T}) \geq k_B, \tag{3.1}$$

where k_B is Boltzmann constant, ΔE and $\Delta(1/T)$, generally, could be some functions of their arguments.

Recent progress in nonequilibrium thermodynamics brought the new limiting condition for total entropy rate in Markov jump processes [21, 22] dealing with the state arbitrary far from an equilibrium, which could be written as

$$\frac{\dot{S}_t}{2k_B} \geq \frac{<J>^2}{<\delta J>^2} \tag{3.2}$$

[21, 22], where $\dot{S}_t$ is total entropy rate, $<J>$ is an average flux, and $<\delta J>$ is variance of flux fluctuations. Further in the text, (3.2) is referred to as *STUR*.

Finally, the interconnection of *RECE* with the classic theorem of Shannon-Hartley [23] on the information capacity of noisy channels is considered. As is known, this theorem claims that theoretically it is possible to transmit data through a noisy channel in errorless mode if the capacity of channel C exceeds the rate of information transmission R, i.e., if

$$C \geq R \tag{3.3}$$

If (3.3) is violated, the transmitted information contains some percentage of errors that could be estimated.

1.3.2. Assumptions

Let internal energy U and temperature T be interrelated, so

$$U = g(T / T_0)k_B T \tag{3.4}$$

where $g = g(T/T_0)$ is an unitless normalized quantity, and T_0 is some reference temperature.

Introduce compressibility factor

$$\Delta X = g\frac{A}{U} \tag{3.5}$$

where A is physical work.

In section 1.3.3, we will believe that OTS is in the compressible state, so

$$dX \sim 1 \tag{3.6.a}$$

while in section 1.3.4, it is in the quasi-incompressible one with

$$dX << 1, \text{ but } \delta X \neq 0 \tag{3.6.b}$$

Further, we will employ energy conservation law in its thermodynamic form

$$\frac{\Delta U}{Q} = 1 - \frac{A}{Q} \tag{3.7}$$

and the Clausius inequality for entropy [24]

$$\delta Q \leq TdS \tag{3.8}$$

As in this chapter all calculations are accomplished based on (3.8), this estimate is valid for the thermodynamic definition of entropy.

At that, we use the same conceptual apparatus but different methods, so for convenience of results, presenting consideration is phased out into separate sections.

In section 1.3.3, the earlier introduced interface-factor x is treated as finite, while in section 1.3.4, factor x is treated as infinitesimal quantity.

1.3.3. Finite Interface-Factor *x*

1.3.3.1. Connection between *RECE* and *STUR*

Employing (3.4), (3.5), (3.7), и (3.8), from (1.6.a)

$$\frac{\Delta S}{k_B \Delta X} \geq \frac{1}{1-\cos(\Delta\varphi)} \tag{3.9}$$

As $1 - cos(\Delta\varphi) = 2sin^2(\Delta\varphi/2) \leq 2sin^2(\Delta\varphi)$, let $sin(\Delta\varphi) = \Delta J/J$, then

$$\frac{dS}{k_B} \geq \frac{\Delta X}{2}\frac{J^2}{\Delta J^2} \tag{3.10}$$

where J is a module of an energy flux vector J and ΔJ is the tangential component of vector J.

At $\Delta X \geq 1$, (3.10) is equivalent to

$$\frac{dS}{k_B} \geq \frac{J^2}{2\Delta J^2} \tag{3.11.a}$$

Consider case $\Delta X \leq 1$. Since from inequality $y \geq ax$ at $a \leq 1$ follows that $x \geq y$, then applying that to (3.10), we obtain the inequality with the opposite sign that using (3.10) could be written as

$$\frac{dS}{k_B} \leq \frac{1}{1-\cos(\Delta\varphi)} \tag{3.11.b}$$

Note that having averaged (3.11.a) on the time interval $\tau \sim \Delta t$ and denoting $\dot{S}t = \Delta S/\Delta t$, we have solution that up to a constant multiple coincides with *STUR* (3.2).

Highlight here that (3.11.b) does not violate (3.10) and, therefore, the Clausius inequality (second thermodynamic law) (3.8).

Also, pay attention that in its physical meaning, quantity ΔX deals with the significance of compression effects (ratio of physical work A and internal energy U) as follows from (3.5). It means that unless condition $\Delta X \geq 1$ $(A \geq U/g)$ compromises the energy integrity of a system, both physical scenarios (3.11.a) and (3.11.b) have equal probability to occur.

In section 3.4, our calculations will be based on (3.11.b).

1.3.3.2. Connection between *RECE* and *TUR*

Write $\Delta U = (\Delta U/\Delta T)\Delta T$, then (1.6.a) could be expressed as

$$\frac{(\Delta U/\Delta T)\Delta T}{Q} = \cos(\Delta\varphi) \tag{3.12}$$

Multiplying (1.6.a) and (3.12) and using (3.8), we have

$$\Delta U \Delta(1/T) \leq \frac{\Delta S^2}{\Delta U/\Delta T}\cos(\Delta\varphi)^2 \tag{3.13}$$

where

$$\Delta U \Delta(1/T) = \frac{\Delta U \Delta T}{T^2}$$

In (3.13), represent $\Delta U/\Delta T$ as

$$\frac{\Delta U}{\Delta T} = \Delta U \Delta(1/T) \cdot \frac{T^2}{\Delta T^2}$$

and find the square root from both parts of the resultant relation, then we obtain

$$\frac{1}{\cos(\Delta\varphi)} \Delta U \Delta(1/T) \cdot \frac{T}{k_B \Delta T} \le \frac{\Delta S}{k_B} \tag{3.14}$$

Using (3.11.b), replace $\Delta S/k_B$ in (3.14), so

$$k_B \ge L \cdot \Delta U \Delta(1/T) \tag{3.15}$$

where dimensionless

$$L = \frac{T}{\Delta T}\left(\frac{1}{\cos(\Delta\varphi)} - 1\right) \tag{3.16}$$

As at $cos(\Delta\varphi) = 1$, $L = 0$, and at $cos(\Delta\varphi) \to 0$, $L \to \infty$, we may think of L as an indicator of how significant the deviation of the actual energy exchange pattern (1.6.a) is from an ideal one (1.6.b).

Then, employing reasoning we did before, we conclude that at $L < 1$, from (3.14) follows *TUR* (3.1).

To avoid confusion, below interface-factor $cos\ \varphi$ will be temporarily designated as $cos\ \beta$. Then, solving inequality $L \le 1$, we obtain that it is valid if at fixed $T/\Delta T$ quantity $cos\ (\Delta\beta) \in [½, 1)$ or extending *cos* to complex plane, $cos\ (\Delta\beta) \in [½, \varphi]$.

So, we showed how to obtain *TUR* (3.1) from (1.6.a) for finite x.

1.3.4. Infinitesimal Interface-Factor *x*

1.3.4.1. Connection between *RECE* and *STUR*

Using (3.7) in its differential form, (3.4), (3.8), and Taylor expansion for $cos(d\varphi)$ [25], eq. (1.6.a) could be written as

$$\beta^2 \frac{dS}{k_B dW} \geq 1, \tag{3.17}$$

where infinitesimal

$$\beta^2 = \sum_{k=1}^{\infty} \frac{(-1)^{k+1}(d\varphi)^{2k}}{(2k)!} \cong \frac{d\varphi^2}{2}$$

Also, in the infinitesimal approximation, $d\varphi \sim sin(d\varphi) \approx dJ/J$, hence from (3.17)

$$\frac{dS}{2k_B dW} \geq \frac{J^2}{dJ^2} \tag{3.18}$$

Then, based on (3.6.b) and the reasoning in Section 1.3.3, we can write for the range $[dX, dX+ d\alpha]$

$$\frac{dS}{k_B} \leq 2\frac{J^2}{dJ^2} = \frac{2}{\sin^2(d\varphi)} \tag{3.19}$$

where $d\alpha \sim dX$.

1.3.4.2. Connection between *RECE* and *TUR*

Following logics of 1.3.3.2, consider the relation between (1.6.a) and *TUR* (3.1). With this aim, use (3.14), (3.19) and replace the differences Δ with the differentials d, then avoid repetition,

$$k_B \geq M \; dU \, d(1/T) \tag{3.20}$$

where dimensionless

$$M = \frac{\sin^2(d\varphi)}{2\cos(d\varphi)} \frac{T}{dT} \tag{3.21}$$

For quantity M, we can repeat the above said for L. Indeed, at $cos(\Delta\varphi) = 1$, $sin(\Delta\varphi) = 0$, and $M = 0$. In the opposite case, at $cos(\Delta\varphi) \to 0$, $sin(\Delta\varphi) \to 1$, and $M \to \infty$. So, like L, quantity M characterizes the extent to which the existing pattern of energy exchange (1.6.a) is close to an ideal case (1.6.b).

So, if $|M| \leq 1$, then from (3.20) we can obtain *TUR* (3.1). Also, for fixed $T/(2\Delta T)$, factor $cos(\Delta\varphi) \in [0, 1/\varphi)$ or, extending *cos* to the complex plane, $cos(\Delta\beta) \in [1/\varphi, \varphi]$.

So, we showed how to relate *TUR* (3.1) to (1.6.a) for infinitesimal x.

1.3.5. Connection with Shannon's Theorem

As was mentioned above, Υ can be interpreted as a volume for phase space M (Figure 1.2). Then in M, the maximum number of possible microstates W at given rate y is $(d\Upsilon/dy)dy$ [33]. Therefore, from (1.12.a) and using Gibbs' definition [28], Shannon's entropy up to a constant additive A_0 is

$$\Delta S = \ln W = \ln|\ln y| + A_0 \tag{3.22}$$

which fully coincides with the main part of differential entropy (1.23.a) discussed below.

Rewrite (1.8.a) in the form

$$\Upsilon = -\iint_M \delta\Upsilon(x, y) = -\int_{-1}^{1} dx \cdot \int_0^y \ln y \, dy \tag{3.23}$$

then we can state that (3.23) declares maximum possible efficiency Υ for energy exchange through interface of *OTS*, which is attained for given random conditions in confidentiality range $[0, y]$.

In other words, (3.23) defines the amount of information that can be transmitted on a noisy channel (through интерфейс *OTS*) at transferred quantity Q and bandwidth y. In this sense, (3.23) directly refers to the context of Shannon's theorem on the interval of confidentiality $[0, y]$.

To make the connection to Shannon's theory even more obvious, rewrite (3.23) as

$$\int_0^1 dy \int_{-1}^1 \delta\Upsilon(x,y)\, dx = \left| \int_1^e dy \int_{-1}^1 \delta\Upsilon(x,y)\, dx \right| = 1 \tag{3.24}$$

where $|z|$ designates modulus of z. Relation (3.24) means that range $0 \le y \le GRP_R$ includes all possible scenarios of energy exchange during one total lifetime cycle of evolution (full history of random changes of $\delta\Upsilon$).

On the other hand, (3.24) shows that to guarantee error-less transmission of energy ($dU = Q$) through interface *OTS*-surroundings, y should change in the range $0 \le y \le SP$ (or $SP \le y \le GRP_R$ for reverse transmission). Abovesaid prompts employing the principles of information theory and calculation of information (Shannon) entropy as in (1.23.a).

1.3.6. Lower Limit for Thermodynamic Entropy

Above, an interplay of Υ with information and differential entropy was shown; however, a possible connection with thermodynamic entropy was not commented on, so the next section deals specifically with this issue.

Earlier, information entropy ΔS^y with a quasi-continuous spectrum was calculated (1.23.a) due to a limitation imposed on quantity H_x.

Analysis of reasons causing the appearance of limitations on the value of entropy was discussed in periodicals [21, 22, 26, 27]. Existence of the localized uncertainty for the small-scale fluctuations in equilibrium thermodynamics and for the large-scale fluctuations arbitrary far from equilibrium for Markovian processes in non-equilibrium thermodynamics was demonstrated in [21] and referenced here to as *STUR*. In [22], at the nonequilibrium steady states of Markovian processes, a few universal bounds valid beyond the Gaussian regime were derived. Authors [27] declared the existence of a new class of thermodynamic uncertainty inequalities, which have revealed that dissipation constrains the fluctuations in steady states arbitrarily far from equilibrium. A possible way to come to the identical constraint for the thermodynamic entropy based on (1.2) and Clausius inequality (3.8) for the physical vicinity was demonstrated in [26] as well as above in 1.3.3 and 1.3.4.

Generalizing the said above in [21, 22, 26, 27], the constraint for thermodynamic entropy could be given as

$$dS \geq \frac{k_B}{2}\frac{1}{y^2} \tag{3.25}$$

where $y = <\Delta J>/<J>$.

So, integrating (3.11.a), obtain for thermodynamic entropy

$$\Delta S \geq -\frac{k_B}{2}\frac{1}{|y|} + C_H \tag{3.26}$$

where C_H is an integration constant.

Assuming C_H is bounded like H_x in (1.32), we obtain that solution (3.26) may also have the quasi-continuous y-spectrum that in its form (shown in Figure 3.1) is close to the depicted in Figure 1.5.

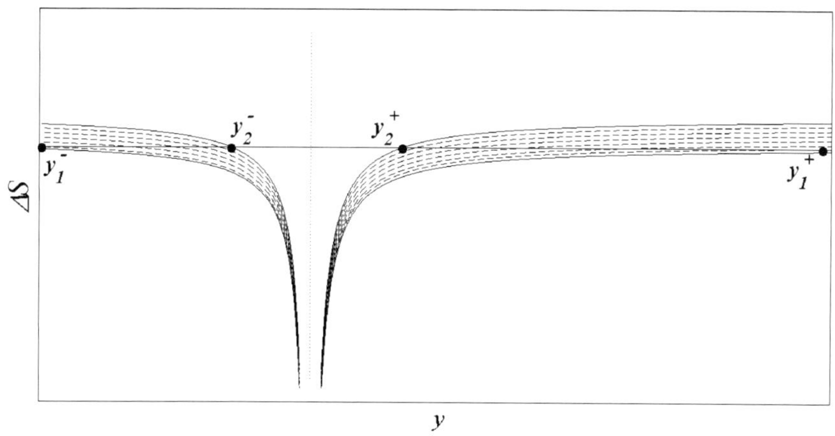

Figure 3.1. Lower threshold for thermodynamic entropy ΔS. In the plot, designation of the horizontal axis is the same as in Figure 1.1, by the ordinate axis, the lower threshold for thermodynamic entropy ΔS is indicated. The meaning of the points $y_1^{\pm}$ and $y_2^{\pm}$ is identical to the y-points in Figure 1.5.

In summary, the lower threshold of thermodynamic entropy demonstrates a close profile of variations with the earlier relations for statistical (1.14) and differential entropy (1.23.a), which provides additional argument in favor of the correctness of the employed approach.

1.3.7. Discussion

The above results indicate that uncertainty of state caused by the necessity of random choice from an infinite number of scenarios of the *OTS* interface, on the one hand, permits making a qualitative estimate for a range of internal features *OTS*, which favors *TUR* (3.15 and 3.20), on the other hand, imposes restrictions on the minimum rate of entropy production in energy exchange, which creates an instrument for optimization of this process. Besides, deep interconnection of *RECE* (1.6.a) to canonic relations of uncertainty (3.1, 3.2) provides a weighty argument in favor of the validity of *RECE* itself. Finally, from a mathematical point of view, borders for using the obtained results (roots of inequalities $| L | \leq 1$ and$| M | \leq 1$) also have a direct connection with the golden ratio, which can be understood as an indication of the existence of this or that sort of dynamic balance (maximum optimality) of the energy exchange process. It is not the only manifestation of a dynamic balance state in energy exchange. As is shown in Chapter 2.2, dynamic balance accompanies the occurrence of the most important points of the discrete spectrum in the evolution of *OTS*.

In more detail, conducted research reveals that combining *RECE* (1.6.a) with Clausius inequality (2nd thermodynamic law) (3.8), we can get the canonical form for *TUR* (3.1) and *STUR* (3.2). The major difference of our approach from the one used before is that we employ the modified *ECE* with the changeable interface factor $x = cos(d\varphi)$. Physically, x is an indicator of the instantaneous pattern of energy exchange flow between *OTS* and its surroundings.

Remind that in chapter 1.1 factor x is taken into consideration at integration, while in this chapter it is maintained. So, generally, we use the same approach through both chapters, the only difference is the details of calculations.

Underline here that while we did obtain the solution family containing canonical *STUR* (2.2), we did not prove in general the validity of *TUR* (2.1). What we did was discover conditions when *TUR* is valid.

The concept of thermodynamic uncertainties was considered in a number of articles. As it is shown in [28], in a limit of very high energies, it looks feasible to introduce into the standard thermodynamic uncertainty relation an additional term proportional to the interior energy that leads to the existence of the lower limit for inverse temperature. In principle, eq. (1.6.a) could also be considered as the result of applying to (1.6.b) some extra uncertainty (energy), which ultimately converts (1.6.b) to (1.6.a) and, finally, comes in as a limitation for entropy.

In [29], authors analyze an ensemble in which energy E, temperature T, and multiplicity N can fluctuate, and with the help of non-extensive statistics, they propose a relation connecting all fluctuating variables that generalizes known Lindhard's *TUR* [30]. Assuming that the non-extensive statistics do not employ traditional Boltzmann-Gibbs statistical mechanics, this result seems promising, especially for the research of complicated and yet unexplored objects like nanostructures.

Results [31] assume that an indeterminacy in the fundamental description enables the use of an information theory framework. In this sense, it looks reasonable to find proofs of uncertainty relations based on the results of information theory. We believe that it is in compliance with our reasoning in Section 3.1 about an "uncertainty" meaning of (1.6.a) in contrast to quite determined (1.6.b). In this sense, the results of sections 1.3.5 and 1.3.6 could be considered a more detailed answer to the highlighted issue.

Comparing the violation conditions for (3.1) and (3.2), we may, in a simplified way, say that *STUR* fails at the modest or small extent of compressibility ΔX (3.6) and small deviations of energy flow ($\Delta J < J$), while *TUR* does it at small ΔT and $\cos(\Delta\varphi) \sim 1$. At great ΔS, (3.1) and (3.2), by and large, can be valid at the same time, while at low ΔS, the validity of *STUR* can break, leaving the validity for (1.1) as an option.

In summary, we conclude that this model can work for the description of the processes characterized by high ΔT, ΔJ, ΔU and small $\cos(d\varphi)$, i.e., for the conditions occurring far away from equilibrium.

There is an important difference between cases (1.6.a) and (1.6.b) from the standpoint of the reason for the occurrence of uncertainty. We mean that in the case (1.6.b), we would rather deal with an equilibrium case, and the thermal fluctuation is the essential reason causing the uncertainty. But, if we use (1.6.a), the driving force for an uncertainty is the permanent reproduction of the non-equilibrium conditions when, at any given moment of time, *OTS* has the unlimited set of options (source-sink pairs) with the non-zero probability to pick up.

Rephrasing the abovesaid, the badly we break (1.6.b), the wider range for uncertainty we should claim. In this sense, the emergence of uncertainty for (1.6.a) could be considered a natural amendment of (1.6.b) to get to the higher accuracy of the results.

Concluding this chapter, underline the existence of a connection between (1.31), (3.2), (3.11.a), and (3.18). To be exact, these relations indicate that the greater the uncertainty of H_x, ΔJ, and δJ, the lower the threshold for the production rate of entropy (*dS*). This effect earlier had been already highlighted by other authors, for example [4]. A probabilistic mechanism for incrementing fluctuations amplitude in the vicinity of maximum *dS/dy* was also reported [32]. The existence of such interaction one more time confirms the logical validity of using *RECE* for the description of stochastic exchange processes in *OTS* away from the state of equilibrium.

Conclusion of Chapter 1.3

1. In non-equilibrium thermodynamics, use of *RECE* permits finding an alternative proof for the existence of a group of inequalities limiting change of entropy (3.2), defining conditions for the existence of thermodynamic uncertainty (3.1), as well as binding *RECE* with the capacity of a noisy channel in Shannon's theorem (2.3).
2. The lower threshold of thermodynamic entropy (2.25) demonstrates a close character of variation with relations obtained earlier for Shannon (information) (1.14) and differential entropy (1.23.a), which provides additional argument in favor of the accuracy of the used approach.
3. In general, model *RECE* shows understandable logical interconnection with conventional relations of nonequilibrium thermodynamics, which verifies the applicability of model *RECE* in the description of dynamics of nonequilibrium *OTS*.

Part 2. The Evolutional Aspects of Stochastic Energy Exchange in *OTS*

Note that, as will be seen in the following sections, in contrast to the discussion being conducted so far, we can no longer withstand the pressure of an evolutionary context of the considered theory. So, that is why we are going to make a formal but important change. This change is about the major relation used before; it is *RECE* (1.6.a). After all, it is just a mathematical system of differential equations with solution (1.12.a), the properties of which were studied above in detail using the classic methods.

However, as will be shown below, the form (1.12.a) is only the high level of solution lying on the surface; the more interesting properties of (1.12.a) are in the heart of its internals. The author goes on a tour to investigate it, and to reflect the change of the guiding line, replace with rare exceptions name *RECE* with *CEL*. We will use the name *CEL* until the end of the book.

Chapter 2.1

The Morphology of Energy Exchange in *OTS*

2.1.1. Evolutionary Triangles

Above, the primary energy relations, the underlying *OTS* model (Chapter 1.1), were established, spectral regularities of evolution and their link to possible scenarios of evolution were shown and substantiated (Chapter 1.2), connections of *RECE* to nonequilibrium thermodynamics and theory of information were traced (Chapter 1.3). In addition, necessary geometry ratios, which are intimately related with energy exchange processes in *OTS*, were presented (App. A). To solve the above-mentioned problems, the classic conceptual basis was applied.

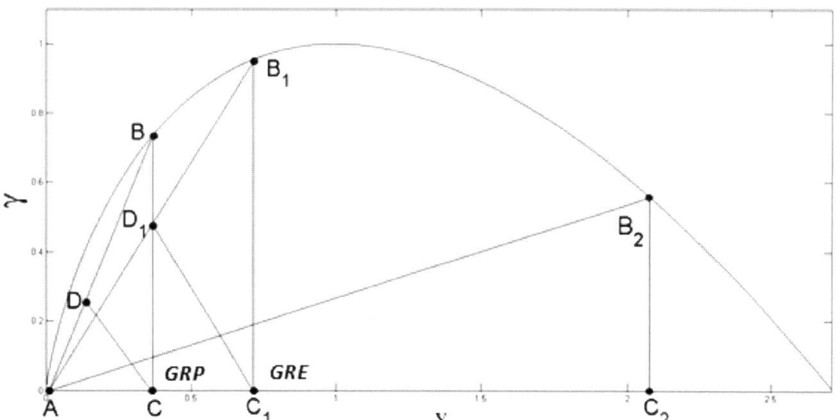

Figure 4.1.a. Evolutionary triangles in the space $y - \Upsilon$. The designation of the axes is the same as in the Figure 1.2. Three evolutionary triangles ABC, AB_1C_1, and AB_2C_2 are shown. These triangles are defined by the legs AC, AC_1, and AC_2 in y and the legs BC, B_1C_1, and B_2C_2, in Υ, appropriately. For ΔABC holds $y = AC$ *(GRP)* and $\Upsilon = BC$ *(2GRP)* as shown in Figure A.1. Such a ratio fits the state of dynamic balance between, on the one hand, the double area of ΔACD ($2E^T$) and, on the other hand, the area of ΔBCD (E^C). Numerical equivalence of energy exchange E and area of ΔAB_1C_1 is shown below. Next ΔAB_1C_1 is defined by leg $y = B_1C_1$ *(GRE)* and $\Upsilon = AC_1$ *(4/3 GRE)*. In point y = *GRE*, the state of dynamic balance between constructive and destructive energy exchange processes is observed. The meaning of the points *GRP* and *GRE* is discussed in detail in the Chapter 2.2.

However, remaining problems, chiefly due to technical and terminological difficulties arising in formalizing description and conclusions, require updating of the employed methods. For these reasons, in particular, the concept of evolutionary triangle (*ET*) is used in further theoretical analysis. Examples of *ET* are shown as triangles ΔAB_nC_n in the phase space $y - \Upsilon$ (Figure 4.1).

So, in this chapter, *ET* will bear the main semantic load at the description of energy exchange in *OTS* evolution.

Further in this chapter, we establish equivalence between k_X, which ties Υ to y (1.12.c), and geometrical k in k-basis (A.2).

Also, the physical meaning of the prominent concept of area of *ET* is disclosed, which permits considering the dynamics of uncovered regularities of energy features in *OTS* evolution.

It should be noted that further we are talking about the triangles, though as likely as it would be possible to talk about evolutionary rectangles or more complicated geometrical figures. However, originally all relationships were established for the triangles, so further we will stick to that approach.

2.1.2. Physical Meaning of Geometrical Results (*K*-Basis)

On one hand, for nonequilibrium energy exchange in *OTS*, as follows from (1.12.b)

$$k_X = \frac{\Upsilon}{y} \tag{4.1}$$

where k_X is defined in accordance with (1.12.c).

On the other hand, from geometry ratio (A.5) in *ET*

$$k = tg \square \ ABC = \frac{BC}{AC} = \frac{\Upsilon}{y} \tag{4.2}$$

Compared (4.1) with (4.2) and accounting (2.13), we come to the conclusion that

$$k = k_X = r \tag{4.3}$$

So, the physical factor of energy exchange k_x (Chapter 1.1), geometric factor of *ET* image k (A.5), and eigenvalue of interface operator $L(y)$ called r (Chapter 1.2) in the considered model are tied; moreover, they are numerically equal. It means that earlier relations containing k, k_X, and r are directly applicable for cross-description and interpretation in the terms of physics, geometry, and linear algebra.

Then, based on established interconnection between different ways for k-description, it makes a lot of sense to unify our description even more. The point is that the majority of mathematical formalisms above use dependence on y and k only. However, as was highlighted above, y and k are tied, for example (1.12.c). So, rid of y and keep dependence on k only. It means that all calculations will be taking place within the k-basis, in which for ΔABC holds $AC = 1$, $BC = k$, and $AB = (k^2 + 1)^{1/2}$ (Figure 4.1).

In part, the area of ΔABC (further K-triangle) in k-basis is

$$S_n = \frac{k_n}{2}$$

from which

$$k_n = 2S_n \tag{4.4.a}$$

Further, all computations will be done in the k-basis.

In agreement with above, clarify the matter of formal equivalence between geometrical and physical description of energy exchange. With this purpose, consider the physical meaning of geometrical area in the discussed theory.

Obviously, in the phase space $y - \Upsilon$, an area of a rectangular triangle is proportional to the product of legs y and Υ. Let us remind the reader that Υ is efficiency of energy exchange, or, in other words, it reveals the rate of energy utilization or leaving *OTS* at given energy flow y. As a result, product $y \cdot \Upsilon$ yields a quantity of energy participating in exchange itself. It is worth noticing that in its meaning, the integral $\int \Upsilon dy$ also is an energy; this matter is discussed in detail below.

Formally, the discussed approach allows an easy extension to the space of three dimensions contributing to formation of the *3-D* figures. The natural third dimension can be the range for random $x = \cos\varphi$ (1.5), which is $[-1, 1]$.

At that transformation, the rectangular triangle becomes the right-angle prism, while a curvilinear triangle the asymmetric semi-cylinder. Then,

$$V_n = S_n \cdot x \tag{4.4.b}$$

In physical meaning, quantity V_n still appeals to an energy referred to as the range x. So, in this sense, an area of the flat geometrical image in the space $y - \Upsilon$ or a volume taken in the space x - $y - \Upsilon$ numerically still mean an exchange energy. Further, replacing the area with suitable energy is carried out by default, except in especially declared cases.

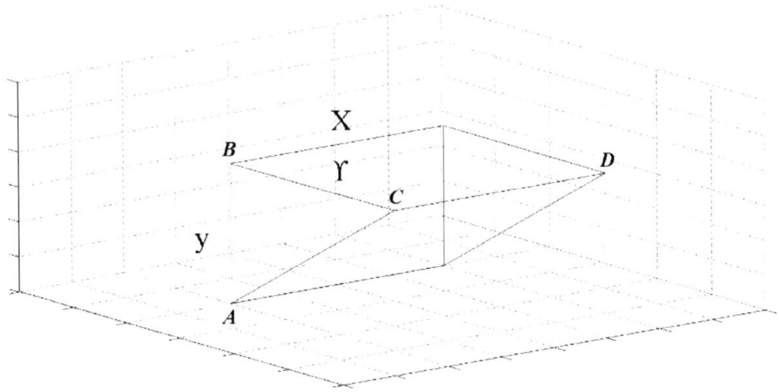

Figure 4.1.b. Right-angle prism in the space x - $y - \Upsilon$. The designation of y and Υ the axes is the same as in the Figure 1.2. The volume is defined by the product *AB, CD,* and *BC* along the axes *x, y,* and Υ.

2.1.3. Energy Balance at *k = 1*

In the previous chapter, we demonstrated how competition between probabilities to hit the interval of "success" (evolutionary scenario) *[0, ln y]* and interval of "failure" (non-evolutionary scenario) *[1, 1 - ln y]* can explain discreteness of Υ and quasi-continuity of entropy ΔS. In this section, we trace the integral transformation of ranges *[0, ln y]* and *[1, 1 - ln y]* for a clear understanding of the role these ranges play in energy exchange. With this purpose, consider Figure 4.2.

Curvilinear $\Delta AFBCA$ is composed by the legs *AC*, *BC*, and arc *AFB*. As arc *AFB* is $\Upsilon(y)$, then area of $\Delta AFBCA$ can be calculated as $\Delta AFBCA$ =

$\int_0^y \Upsilon(y)dy = \int_o^y (y - y\ln y)\,\mathrm{dy}$ as per (1.12.d). So, the area of *ΔAFBCA* is numerically equal to the portion of energy circulating through the *OTS* interface and tied with real microstates in the range *[0, ln y]*, call this energy the total energy *T*. In other words, area of *ΔAFBCA* is integral image of quantity *ln y*.

Curvilinear figure *ADCA* is composed of leg *AC* and arc *ADC*. It is easy to show by direct integration that arc *ADC* is $\int_0^y (1 - \ln y)dy = y\ln y.$. Then area *ADCA* can be calculated as $\int_o^y y\ln y\,\mathrm{dy}.$ So, area of *ADCA* is numerically equal to portion of energy circulating through the *OTS* interface and tied with dropping microstates in the range *[1, 1 - ln y]*, call this dropping energy *W*. In other words, the area of *ADCA* is the integral image of quantity *1 – ln y*.

Energy meaning of area of rectangular Δ*ABC* is defined later.

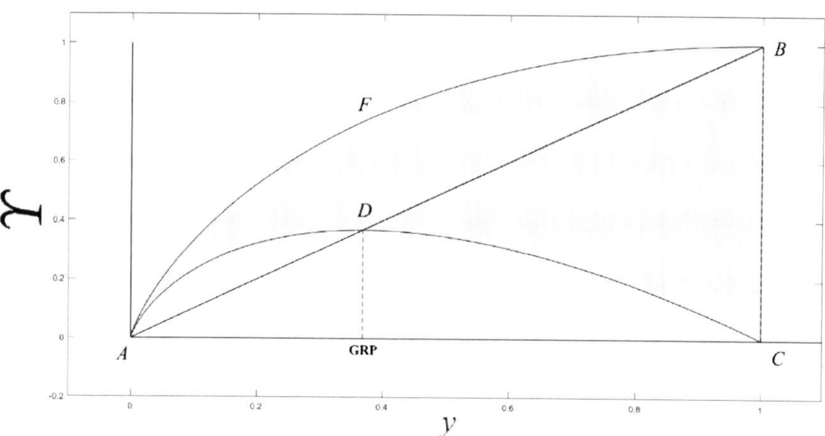

Figure 4.2. Dependence an efficiency of energy exchange *Υ* rate on rate *y* at *k = 1*. The designation of the axes is the same as in Figure 1.2. Curvilinear *ΔAFBCA*, defined by legs *AC*, *BC*, and arc *AFB*, has area numerally equal to the total exchange energy *T* circulating through interface *OTS*. Curvilinear figure *ADCA*, defined by leg *AC* and arc *ADC*, has area numerally equal to exchange energy *W* circulating through interface *OTS* for the dropping microstates. As area *ADCA* is equal to area of elliptic sector *AFBDA*, then area *ΔABC* is difference of areas *ΔAFBCA* and *AFBDA*, and for considering case *k = 1* characterizes the transformation (evolutionary) energy. For visibility, the point *y = GRP* is shown (for details, see Chapter 2.2).

Now, compare the area of $AFBA$ and $ADCA$. From Figure 4.3 follows that, on one hand,

$$S_{AFBCA} = S_{ABC} + S_{ADCA} \tag{4.5}$$

On the other hand,

$$S_{AFBCA} = S_{ABC} + S_{AFBA} \tag{4.6}$$

so, S_{ADCA} (elliptic sector) = S_{AFBA} = $\int y \ln y dy$.

Also, from Figure 4.3, it follows that as $dT/dy \neq dW/dy$, there exists non-zero energy

$$\frac{dE}{dy} = \frac{dT}{dy} - \frac{dW}{dy}$$

So, finally, at $k = 1$ holds

$$\frac{dT}{dy} = \frac{dE}{dy} + \frac{dW}{dy} \tag{4.7}$$

with the following physical meaning.

An energy T is the total exchange energy. Energy W is the part of total exchange energy that originally was accumulated in real microstates of exchange, and later, at the achievement of Υ of discrete domain area (Figure 2.3 (a)), it was released at the conversion of real microstates to imaginary. We tentatively call W a dropping energy, assuming that at conversion, ultimately, this energy moves to the loss cone [4] as heat. Analogue with a loss cone looks suitable due to imposed limitation on $x = \cos \varphi$ at $y > GRP_L$ (section 1.2).

Then, E is the difference between total and dropping energy. In other words, in (4.7), E is an energy to which T converges at the expense of dropping energy W (Figure 4.3).

$$E(y_n) = \lim_{W \to 0} T(y) \tag{4.8}$$

In this sense, E in Figure 4.3 is the exact solution of system (1.2). Also, from Figure 4.3,

$$E = 0.5y^2 \tag{4.9}$$

and, therefore,

$$T = W + 0.5y^2 \tag{4.10}$$

We return to discussion on the meaning of energy E in the next section.

2.1.4. Energy Balance at $k \neq 1$

2.1.4.1. Geometrical Method

Consider the more general case when $k \neq 1$ (Figure 4.4).

Let us write down the expression for total area $\Delta AJHDFA$ in the usual kind of sum of area for elliptic sector $AHGA$ and triangle AHF

$$T = S_{AHGA} + S_{AHF} = S_{AHGA} + \frac{1}{2}ky^2 \tag{4.11}$$

On the other hand, from (4.11), the same area can be expressed as

$$T = S_{\Delta ADF} + W_{new} = 0.5y^2 + W_{new} \tag{4.12}$$

where W_{new} is the new value of dropping energy. Equating (4.11) and (4.12), then

$$S_{AHGA} = W_{new} + 0.5y^2(1-k) \tag{4.13}$$

In turn, from (4.13) follows that W_{new} is

$$W_{new} = S_{AHGA} + 0.5y^2(k-1) \tag{4.14}$$

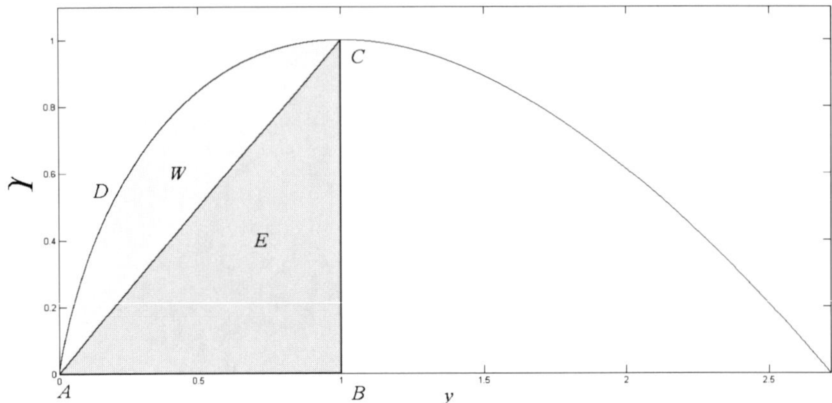

Figure 4.3. The meaning of evolutionary energy E as limit of total exchange energy T at $W \to 0$ for $k = 1$. Designations of the axes are the same as in Figure 4.2. Total exchange energy $T = E + W$. Dropping energy W is shown in light grey, evolutionary energy E in dark grey.

The difference between (4.12) and (4.10) is that at $k \neq 1$, the area of elliptic sector $AJHGA$ is not equal any longer to area $\int y \ln y dy$ (dropping energy W), and, as a result, area ΔAHF partly includes a portion of dropping energy ΔW (ΔAHD) shown in light grey. This sign-changing addition to W can be written as

$$\Delta W = 0.5 y^2 (k-1) \qquad (4.15)$$

or

$$W_{new} = W_{k=1} + \Delta W \qquad (4.16)$$

where $W_{k=1}$ is dropping energy at $k = 1$. Graph for y-dependence of dropping energy is shown in Figure 4.6.

In Chapter 2.3, it will be shown that the main role in the mechanism of energy development of OTS plays the energy E determined by area ΔAHF in Figure 4.4. Dynamic partitioning of area ΔAHF (energy E) in CEL is critical for the existence of competition between energy factors of inheritance and variability that is directly bound with the driving force of energy evolution. It explains the introduction of the term "evolutionary energy" for area (energy) ΔAHF.

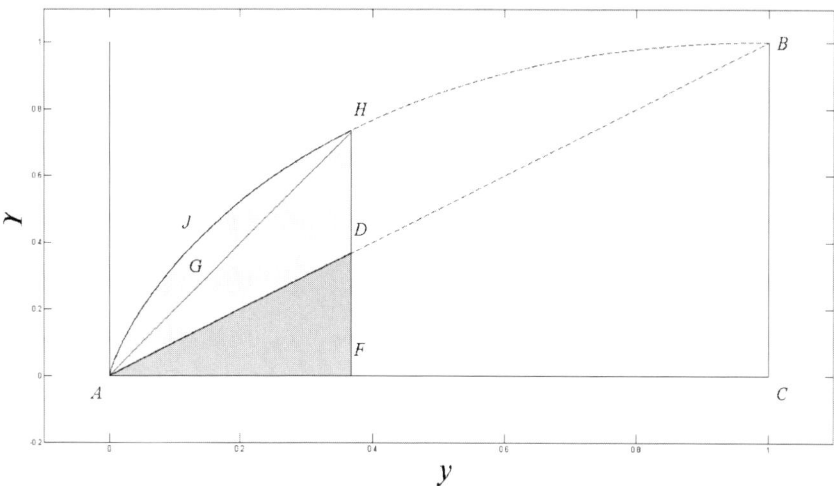

Figure 4.4. Forming of addition to dropping energy *W* at *k ≠ 1*. The *y*-dependencies of exchange energy are shown. The designations of the axes are the same as in Figure 4.3. Area of curvilinear *ΔAJHDFA* with legs *AF*, *HF*, and arc *AJH* is the total energy *T*. Area curvilinear figure *AJHDA* (shown in light grey) with legs *AD*, *HD*, and arc *AJH* is the dropping energy *W*. New value of *W* is not equal to area of elliptic sector *AJHGA*, i.e., $S_{\Delta AJHDFA} - S_{AJHGA} \neq S_{\Delta AHF}$. The rest of energy for ordering and reconstruction of energy exchange is shown in dark grey *ΔADF*. Area *ΔAHF* with legs *AF*, *HF*, and *AH* is an evolutionary energy *E* (in detail see Chapter 2.3). Line *ADB* and arc *AJHB* demonstrate relative position for *k = 1*.

2.1.4.2. Analytical Method

Look at the energy balance of *OTS* at *k ≠ 1* from an analytical point of view.

As before, a total energy circulating through the interface of *OTS* numerically equals the area under *Υ(y)* (aggregation of grey and white domains in Figure 4.2).

Based on (1.12.d), the dropping energy is

$$W = \int_0^y (\Upsilon(z) - z)dz = \frac{1}{4}y^2(1 - 2\ln y) \tag{4.17.a}$$

with the rate

$$\frac{dW}{dy} = y\ln y \tag{4.17.b}$$

Highlight that (4.17.a,b) applies to the phase at $GRP_L \leq y \leq GRP_R$ only, as comparison of *lny* with *1* makes any sense only within this range.

Then, difference

$$E(y) = T(y) - W(y) = \frac{1}{2}ky^2 \tag{4.18.a}$$

and rate

$$\frac{dE}{dy} = \frac{1}{2}y - y\ln y \tag{4.18.b}$$

where

$$k = \frac{\Upsilon_n}{y_n} = k_n = 1 \pm \ln y_n = 1 \pm \frac{1}{n} \tag{4.19}$$

As before for the case $k = 1$, relation (4.18.a,b) indicates that the difference *E* between total energy *T* and dropping energy *W* for evolving *OTS* cannot be zero, as some energy should remain and be used by *OTS* for its transformation (Figure 4.5). Below, we show that *E* is a major resource of changes in *OTS* on the stage of discrete spectrum. Plot for the rate of energy *E* is shown in Figure 4.6.

So, the equation of energy balance for *OTC* in full agreement with (4.7) can be rewritten as

$$\frac{dT}{dy} = \frac{dE}{dy} + \frac{dW}{dy}$$

bearing in mind that now, in contrast to (4.7), terms *E* and *W* experience regular changes.

From (1.12.d, 4.17, 4.18) follows that *dE/dy* meets

$$\frac{dE}{dy} = \frac{1}{2}\left(\frac{dT}{dy} + \frac{dW}{dy}\right) \tag{4.20}$$

for each single y-point. In other words, for each y in the range *[0, 1]* holds.

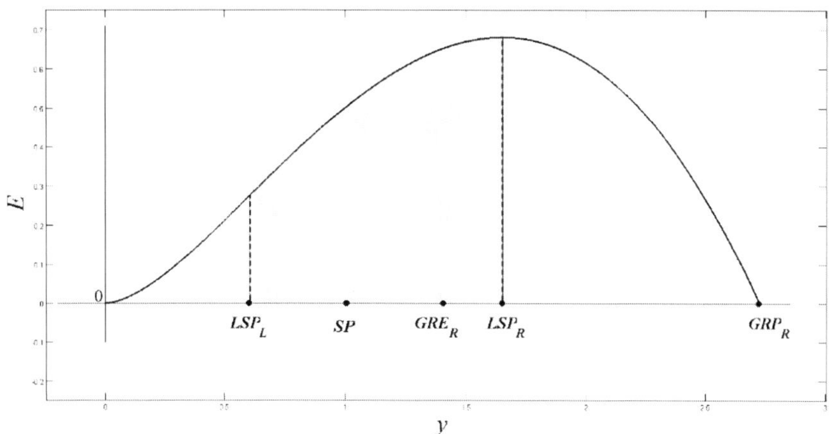

Figure 4.5. In the plot, the dependence of evolutionary energy E on energy rate y is shown.

By an abscissa axis, energy rate $y = J/J0$ is indicated, and by an ordinate axis, evolutionary energy E. The bifurcation points LSP_L, LSP_R, SP, and GRP_R are shown. The evolutionary range between the points LSP_L and LSP_R is shown in light grey. Maximum energy E is at point LSP_R. Point LSP_L is an inflection point for curve $E(y)$, while SP is a point of stationarity.

$$\begin{aligned}&(\frac{dT}{dy}-\frac{dW}{dy}):(\frac{dE}{dy}-\frac{dW}{dy})=2\\&(\frac{dT}{dy}-\frac{dW}{dy}):(\frac{dT}{dy}-\frac{dE}{dy}-)=2\end{aligned} \tag{4.21}$$

Then, based on the theorem (A.1), the energies $T - W$, E - W, and T - E are in the state of dynamic balance at any point of the *[0, GRP_R]* range [5]. So, if the energy profile of *OTS* lies in the vicinity of curves (4.21), it creates the most favorable conditions for the evolutionary process. At an opposite sign of dW/dy, the dynamic balance moves to the point $y = GRE_R$. Also, notice that at $y > LSP_R$ the inequality $\Upsilon < dW/dy$ changes its sign (Figure 4.6.a,b). Formally,

(4.20) can be extended on interval *[1, GRP_R]*, however, it does not make sense as *dW/dy* becomes negative (Figure 4.6.b).

Above ratios assume simple physical interpretation. At given *y*, the difference *dT/dy* – *dW/dy* forms an available energy "corridor," within which the most effective energy transformation of *OTS* takes the equidistant trajectory from the trajectories *dT/dy* and *dW/dy* (Figure 4.6.a,b). Since significant fluctuation background eminently presents on all stages of *OTS* development, then the real *OTS* trajectory remains random, where presented deterministic regularities play the role of energy infrastructure for evolution. The existence of a dynamic balance curve across the entire range *[0, GRP_R]* shows that evolution is a natural way for *OTS* development within this range.

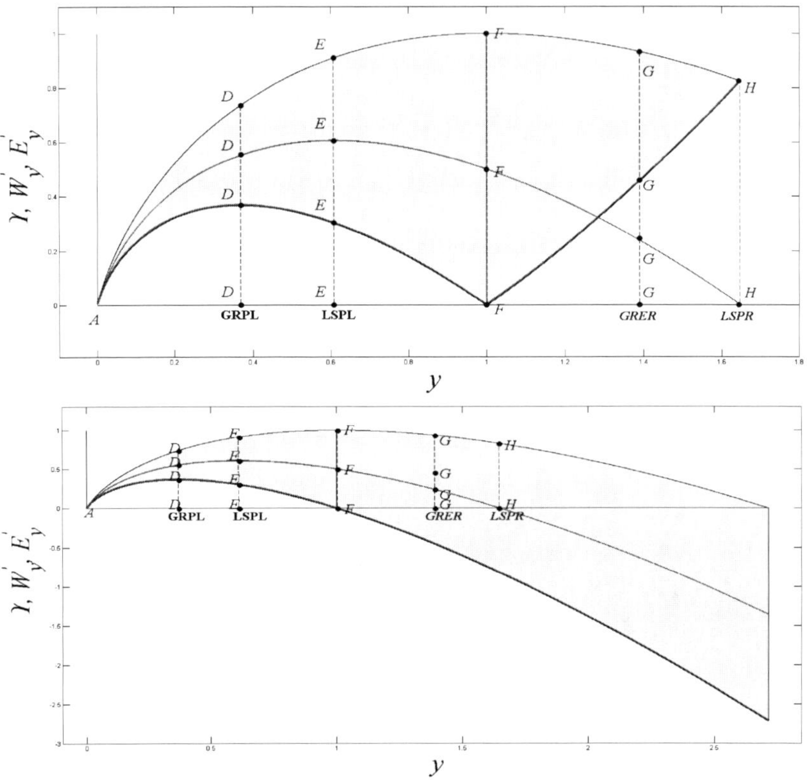

Figure 4.6. a,b. The production rate for *y*-dependence of total $Y = dT/dy$, dropping *dW/dy*, and evolutionary *dE/dy* are shown. Dynamic balance between available *dY/dy* – *dW/dy* and each *dE/dy* – *dW/dy*, *dY/dy* – *dE/dy* is kept in each point *dE/dy* at $0 \leq y \leq GRP_R$ (lower panel). Sign *dW/dy* in the upper panel changed to the opposite for visibility.

Conclusion of Chapter 2.1

1. Physical factors of energy exchange k_x (chapter 1.1), geometric factors of evolutionary triangle k (App. A), and eigenvalues r of evolution operator $L(y)$ (chapter 1.2) in the considered model are tied; moreover, they are numerically equal. It means that earlier relations containing k, k_X and r, are directly applicable for cross-description and interpretation in the terms of physics, geometry, and linear algebra.
2. From a large-scale standpoint, the energy development of *OTS* can be thought of as moving along the chain of consecutive states of dynamic balance.
3. In this model, the energy resource of evolution is the dynamically changing part of total energy exchange E (evolutionary energy).
4. The rate of total exchange energy dT/dy breaks up into two parts, the rate of evolutionary energy dE/dy and dropping energy dW/dy, i.e., $dT/dy = dE/dy + dW/dy$.
5. In this model, an area numerically equals an exchange energy.
6. Geometrically, in *2-D* space, total exchange energy is an area of a curvilinear triangle defined by legs y (rate of energy exchange) and Υ (efficiency of energy exchange), whereas energy E is an area of a right-angled triangle defined by the same legs.
7. The considering approach allows an easy extension to the space of three dimensions, contributing to the formation of the *3-D* figures.

Chapter 2.2

The Morphology of *OTS* Evolution

In this chapter, the role of nodes in the discrete spectrum of energy evolution of *OTS* (2.3) as bifurcation points and their properties will be considered. The phenomenon of the appearance of evolutionary phases (along the y-axis) of dissimilar physical meaning that are separated by mathematically clearly identified points of solution bifurcation will be analyzed. From this point of view, it will be suggested to consider the bifurcation points altogether with the suitable evolutionary phases as an energy infrastructure of evolution.

2.2.1. Spectral Harmonics as Points of Bifurcation

2.2.1.1. Point *GRP* (Golden Ratio Point)

In Figure 4.1, evolutionary triangle (*ET*) ΔABC defined by legs $[0, GRP_L]$ in axis y and $[0, 2GRP_L]$ in axis Υ is shown. Relation $\Upsilon(GRP_L) = 2GRP_L$ directly follows from (1.12.a). Then $k = \Upsilon / y = 2$, from which, in agreement with (A.1), suitable parts of area in ΔABC (exchange energy) are bound by golden ratio [1] that can be written as

$$\left(\frac{E+E^T}{2E^T}\right)^p = \left(\frac{2E^T}{E-E^T}\right)^p = \varphi^p, \tag{5.1}$$

Based on (5.1), further, point $y = e^{-1}$, which is a fundamental harmonic y_1 of the evolutionary spectrum (2.3), shall be known as the *Golden Ratio Point*, or briefly *GRP*. Following existing science understanding, in *GRP* the quantities $E - E^T$ and $2E^T$, as well as $E + E^T$ and $2E^T$, are in the state of dynamic equilibrium [2].

For consistency, we extend coverage of the name *GRP* to the paired point $y = e$, though in the latter the ratio (5.1) does not work. In the majority of cases, the acronym *GRP* will appear together with subscripts L or R, i.e., GRP_L or GRP_R, where L refers to the point $y = e^{-1}$, whereas R calls the point $y = e$.

Besides, *GRP* has the following properties. In the point GRP_L:

a) Fundamental harmonic of discrete energy spectrum is observed.
b) Form of spectrum for key evolutionary parameters in *OTS* alternates (in integral efficiency *ϒ*, entropy *S*, eigenvector *k*) between discrete and quasi-continuous.
c) At $H_x = 0$, it is the root of entropy *S* both for the probabilistic (1.14) and differential (1.23.a) definitions.
d) At $H_x = 0$, the rate of entropy production *dS/dy* is maximum both for the probabilistic (1.14) and differential (1.23.a) definitions.
e) Probabilistic dominance of $y = y_{in}$ меняется на $y = y_{out}$ (1.30 a, b).
f) The probability of fluctuating behavior of entropy grows drastically ([20]). This is thanks to fluctuating behavior of ratio $P(y_{out})/P(y_{in})$, where *P* is probability for realization event y_{out} or y_{in}.
g) Eigenvalues of an evolutionary operator *L(y)* acquire the imaginary part (1.2.5).
h) Production of dropping thermal energy attains a maximum by absolute value.
i) Dropping and total energy exchange are in dynamic balance due to the ratio of appropriate production rates (Figure 4.2).

In point GRP_R:

a) maximum root of entropy *S* (1.14) is observed shown in Figure 1.5.
b) maximum of total exchange energy *T* is observed (1.12.d) shown in Figure 2.2.

So, though GRP_R, certainly, is one of the critical points of evolution, at the same time, it does not possess so many outstanding features compared to GRP_L. It suggests the idea that in course of energy development, the scope of differences between the spectral harmonics is reduced. In this sense, the role of GRP_R, ultimately, is to signal the termination of the evolutionary cycle for total production rate *ϒ*.

Notice that GRP_L separates two qualitatively different in its physical essence *y*-domains, and at this point transfers from one operational mode of evolution to another as well as from continuous spectrum to discrete one happens. Above differences (a) – (i) give grounds to classify GRP_L as point of bifurcation. Henceforth, the abovementioned two *y*-domains with opposite properties will further refer to agenesis ($y < GRP_L$) and genesis ($GRP_L \leq y \leq$

GRP_R). Notice that, as follows from Figure 2.2, formally, the range $GRP_R \leq y \leq TP$ (post-genesis) may be considered a single terminal stage of the evolutionary spectrum.

Key properties of GRP_L are schematically presented in Figure 5.1 below.

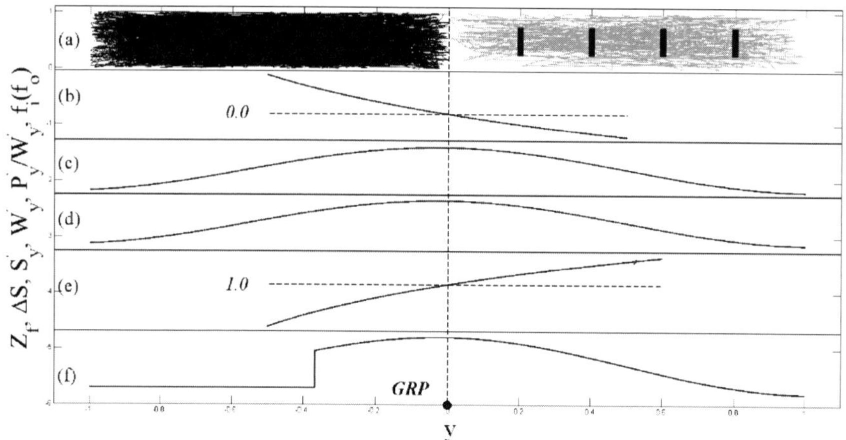

Figure 5.1. Schematic dependence of key parameters of *OTS* at passing through GRP_L on energy exchange rate y. (a) Form of spectrum Z_f is continuous on the left panel and discrete on the right panel; discrete harmonics are shown as thick vertical segments; (b) change of entropy ΔS from positive to negative; (c) rate of entropy production dS/dy is at maximum; (d) rate of unstructured dropping energy production dW/dy is at maximum; (e) rate for evolutionary energy dE/dy matches dW/dy; (f) $f_i(f_o)$ - probability density f_i of energy flux towards *OTS* on probability density for flux f_o from *OTS* out. All quantities are dimensionless. In the vertical axis, notation F'_y denotes dF/dy.

Now, it makes sense to return to discussion on initialization of probabilistic anisotropy of interface $A^y(x)$ in (1.28), where stage of agenesis is bound to predominant probability of energy flows towards *OTS*, while probabilistic directivity of energy flows on the stage of genesis is opposite. As flows of "raw" energy towards *OTS* coincide with growth of activity for stochastic processes, while inside-out flow with growth of ordering, then it would be logical to admit that agenesis is bound with prevalence of random noise component, whereas genesis with processes supporting energy ordering and structuring in *OTS*. It is seen that GRP_L is a horizontal coordinate for maximum of $dW/dy = -y\ln y$. Then, at $y < GRP_L$

$$\frac{dT}{dy}:\frac{dW}{dy}<2 \quad (5.2.a)$$

which breaks R_E, energy flow is highly turbulent, and chaotic regime will dominate in exchange process. At $y > GRP_L$, the opposite pattern will be observed

$$\frac{dT}{dy}:\frac{dW}{dy}>2 \quad (5.2.b)$$

and the ordering will prevail in energy exchange processes. That is why high *dW/dy* is a characteristic feature of agenesis and the post-genesis, i.e., where ordering of energy exchange either has not completed yet or has already degraded to a significant degree.

So, we come to the conclusion that at $y > GRP_L$, the changes of dropping energy play a relatively less role in the energy development of *OTS*.

2.2.1.2. Point of *LSP* (Least Scenario Probable)

Present discovered properties of points $y = y_2 = e^{\pm 1/2}$ in the same format as above for *GRP*. Then, at $y = y_2$:

a) A second harmonic of the discrete energy spectrum is observed.
b) The greatest (for $GRP_L \leq y \leq SP$) and the least (for $SP \leq y \leq GRP_R$) root of differential entropy (1.34) is observed (Figure 1.5).
c) As follows from (2.11), there exists parity between the probability of evolutionary and non-evolutionary scenarios in the energy development of *OTS*

$$\vartheta(n=2)=\frac{1-|\ln y|}{|\ln y|}=1 \quad (5.3)$$

d) Variance for Bernoulli distribution attains its maximum (Figure 2.4).
e) Maximum production rate of evolutionary energy (*dE/dy*) is observed.
f) Maximum of evolutionary energy (in LSP_R) is observed.
g) Production rate of *dE/dy* becomes negative (at $y > LSP_R$).

The above-listed properties eloquently demonstrate the boundary meaning of *LSP*. For example, look at the variance in Bernoulli distribution $D = \ln y\,(1 - \ln y)$. The obvious conclusion is that at $y = LSP$ there is a maximum uncertainty in the results of a random trial to find evolution in the interval of "success" or "failure." It assumes that in *LSP*, the probabilities to discover *OTS* in the *y*-range of success (exchange ordering scenario) and failure (exchange chaotization scenario) match each other making (5.3) a condition of static parity between suitable probabilities.

It should be stressed that (2.7) is only a probabilistic statement of the fact that, to the right of *LSP*, the realization of the evolutionary scenarios is less likely compared to the non-evolutionary. Above explains why in evolution, spectral node $y = y_2$ can be named the point of least probable scenario (*Least Scenario Probable*), or shortly *LSP*. However, in reality, the evolutionary scenarios at $y > LSP$ are also possible. In the next paragraph, we will be considering another probabilistic criterion for the development of evolutionary scenarios with clear physical meaning. This criterion, in part, will deal with one more point of bifurcation: $y = y_3 = GRE$.

In conclusion, notice that the point LSP_R as per Figure 4.6, comes as complete threshold for realization of any evolutionary scenario because in this point an inequality $dT/dy > dW/dy$ changes its sign to the opposite, thereby nullifying any chance for an ordering. This conclusion is also confirmed by the fact of nullifying of an evolutionary energy, i.e., $E\,(LSP_R) = 0$, the curve dE/dy crosses the *y*-axis and becomes negative (Figure 4.6.a,b).

2.2.1.3. Point of *GRE* (Golden Ratio Exchange)

The following point is $y = y_3$ (or as discussed below *GRE*). Its properties are below:

a) It is the third harmonic of the discrete energy spectrum.
b) It is the point where

$$\vartheta(n=3) = \frac{1-|\ln y|}{|\ln y|} = 2, \qquad (5.4)$$

as it directly stems from (2.11). It means that the product of the probability to fall to range $1 - |ln\, y|$ (for dropping energy dW/dy) and to range $|ln\, y|$ (for total energy dT/dy) in this point are in the state of dynamic balance.

c) In GRE_R, the production rates of total and dropping energy

$$\frac{dT}{dy}:\frac{dW}{dy}=2 \tag{5.5}$$

It means that the appropriate energies, the total T and dropping W, are in a state of dynamic balance. Besides, the ratio $R_E = dT/dy: dW/dy$ shares the meaning of the pattern shown in the Table 5.1. It confirms that R_E as a ratio of total (kinetic) and dropping (dissipating) energy plays the role of the Reynolds number [23], which in its extended meaning represents a control parameter that expresses a balance between injected and dissipated energy flows for an open boundary system.

In the Table. 5.1, the highest value of $R_E = 4$ is at $y = GRE_L$. Employing the casual meaning of R_E it may indicate the critical value of the ratio between kinetic and dissipated energy flow at this point, i.e., its border essence.

In contrast to the usual meaning of a critical Reynolds number as a constant quantity, in the course of energy evolution value R_E experiences changes, i.e., it is a dynamic quantity (Table 5.1).

Table 5.1. Critical R_E for energy flow in *SEE*. The indicator of the switch is an integer value of the $R_E = dT/dy{:}dW/dy$ (Reynolds number). The high R_E {2, 3, 4} deal with the *y*-range ($GRP_L \leq y \leq GRE_L$) of constructive mode of *SEE*. The low R_E {0, 1, 2} in the *y*-range ($GRE_R \leq y \leq GRP_R$) rather correspond to the destructive mode of *SEE*

R_E ranges of constructive and destructive *SEE*	
Point	R_E, $dT/dy{:}dW/dy$
GRP_L	2
LSP_L	3
GRE_L	4 (maximum)
GRP_R	0
LSP_R	1
GRE_R	2

Then, in this context, the low R_E signifies the essential energy dissipating (turbulent) component, which does not support ordering at in the total energy flow, while the high R_E assumes the low dissipating (laminar) component creating favorable conditions for self-organizing in total energy flow.

To clarify the meaning of (5.5), resolve it directly. Choose $|lny| < 0$, then $dW/dy = -ylny$, and (5.5) reduces to $y\text{-}ylny\text{=-}2ylny$ with the root $y = GRP_L$. On the other hand, at $|lny| > 0$ rate $dW/dy = ylny$, and (5.5) reduces to $y\text{-}ylny\text{=}2ylny$ with the root $y = GRE_R$. It clearly confirms the meaning of the range $[GRP_L, GRE_L]$ as an area of the higher constructive energy exchange compared to the remaining part of the prime range $[0, GRP_R]$.

It is valuable to note that absence of evolutionary potential in the vicinity of the GRP_R, GRE_R does not remove a phenomenon of energy phasing considered later in section 2.2.2.

2.2.2. More Arguments to Border Meaning of *GRE*

For detailed elaboration of role $y = GRE$, turn to Figure 5.2, where the right triangle, formed by legs $|ln\ y|$ and $1 - |ln\ y|$, is shown.

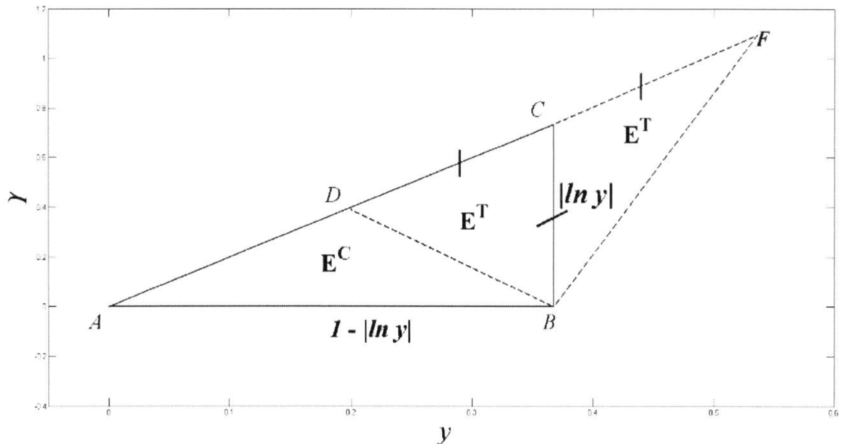

Figure 5.2. Right-angle triangle composed by the legs $|ln\ y|$ and $1 - |ln\ y|$. Shown triangles *ABC*, *AFC* (with area $E^F = E^C + E$), *ADC* (with area E^C), *CDB* (with area E^T), and *CBF* (with area E^T). In ΔAFC, at $\vartheta = 2^{\pm 1}$ ratio (A.1) holds. Equal segments *DB, BF,* and *BC* are marked with a short, thick dash.

As follows from (A.2) and (5.5), there exists a dynamic balance between E^C and E^T. At that, contribution of E^C and E^T to evolution of *OTS* is not the same, which comes from theorem in (A.5), i.e., $Var\ (E^T) \neq Var\ (E^C)$, where *Var* designates dispersion. So, in formal notation at $y = GRE$, quantities $Var\ E^C\ (GRE)$ and $Var\ E^T\ (GRE)$ are in a dynamic balance. It provides a lower threshold for the necessary balance between stability and variability in the process of energy exchange, which is violated at $y > GRE$.

In more detail, earlier it was established that *LSP* marks an equiprobable state of evolution or, in other words, static equilibrium between the parts of evolutionary uncertainty (Figure 2.4), which is manifested in maximum Bernoulli variance. In this sense, at $y > LSP$, chances of non-evolutionary development exceed those of evolutionary scenario, but it does not nullify the probability for evolutionary development completely.

In other words, GRE_l marks a state of dynamic balance between components of evolutionary uncertainty. It means that at $y < GRE_L$, the non-structured dropping energy, albeit grows, but its contribution to proper dominating in processes of energy exchange is still not sufficient compared with the contribution of the total energy *T*. However, at $y > GRE_L$, the contribution of dropping energy is becoming big enough to suppress ordering processes. In this sense, GRE_L is the stronger limitation of chances for evolutionary development compared with LSP_L. As a result, at $y > GRE_L$, the probability of an ordinary system regime of receiving and giving back energy is significantly higher.

So, *GRE*, as well as *GRP* and *LSP*, sets probabilistic borders between physically dissimilar stages of evolution. Based on the above discussion, below the spectral node *GRE* will be called the point of dynamic balance of exchange (*Golden Ratio Exchange*). This *y*-node is one more point of bifurcation, where qualitative change in evolutionary parameters of *OTS* occurs.

2.2.3. Infrastructure of *OTS* Energy Exchange

What is the extent to which model *CEL* is equivalent to *OTS*? Firstly, the *CEL* model accounts for not only the default heat transfer but also the enthalpy flow via mass transfer at the assumption of a permeable boundary. Secondly, *CEL* model accounts for, at least some, work transfer by the transport of charge, momentum, angular momentum, and mass. General arguments in favor of the actual openness of the *CEL* model were also presented above. However, we are not in a position to claim the identity of *CEL* and *OTS*. What we can say with confidence is that the *CEL* model is some approximation of *OTS* indeed. The degree to which this approximation is close to real *OTS* is subject to further discussion.

Conducted research raises an important issue about the existence in the energy spectrum of *OTS* the number of points with the property of boundary between stages of development, in which there are qualitatively different regimes for the functioning of *OTS*. Based on the formal meaning of (2.3), one may say about the existence of an infinite number of evolutionary phases, where points $y = y_n$ play a role as markers.

At that, physically, the qualitative leaps at $y = y_n$ are not identical. If for *GRP* we see a diversity of effects (a – i), in *LSP* the effects being essential nevertheless are not so radical. The trend to reduction of the spectral nodes is also visible for *GRE*, where, actually, there are only two, albeit crucial effects. The presence of any unusual behavior for harmonics with $n > 3$ was not found.

The discovered trend confirms previous conclusion that criticality of changes in y_n decreases with growth n, gradually coming to naught. Similar regularity looks sufficiently logical, accounting that the amplitude of changes in macrostates of *OTS* falls off with y-distance between y_n and *SP*, *i. e.*, with growth n.

Notice that necessity to support certain distance of y_n from *SP* in this model is associated with losing uniqueness for spectral nodes as y_n get closer to *SP*. This is due to the shortening of distance separating individual y-моды (separating distance converges to zero at $n \to \infty$), still supporting a high level of copying accuracy for form (as follows from App. A) and value of energy. It factually makes the discrete y-modes indiscernible.

If we examine the mechanism for the forming of noted drastic qualitative changes, it is possible to conclude that ultimately the line of intercoupling effects goes through a change in probabilistic transparency of the interface (1.28) for energy flow y_{out}. In this sense, the obtained results support position [6-8] that the built-in capacity of *OTS* to adapt through dissipation of redundant energy in the form of suppression of high fluctuations, which in turn is governed by flow y_{out}.

Existence in vicinity of *GRP* a whole series of dramatic changes in behavior of operating parameters indicates that going through this point, structure of energy exchange in *OTS* experiences deep reconstruction manifesting in occurrence of qualitatively new features. In this capacity, the stage of genesis logically and reasonably follows the stage of agenesis. One seems quite rational to believe that the evolutionary essence of agenesis is to stock up (a) a sufficient amount of non-ordered energy (continuous spectrum) and (b) rich prehistory (multitude of closed contours $\delta\Upsilon$) for subsequent effective support of more regulated (discrete spectrum) operation of *OTS* in genesis. Perhaps, by accounting for the obvious concomitant change in evolutionary topology (alternation in type of spectrum), *GRP* could be named the point of the topological phase transition as discussed in [9, 10].

Then, the difference between agenesis and genesis in this model is conditioned by the fact that in agenesis, exchange properties are not tied or tied weakly, which implies practical equiprobability in realization of its microstates. As a result, parameters of the exchange process acquire a continuous spectrum.

It is necessary to notice that the distinctive meaning of *GRP* repeatedly attracted the attention of researchers in the past. One example is the so-called problem of interruption of choosing at verification of only $\sim n/e$ values of sample instead of full sample volume n. It indicates that $1/e$ portion of the trials of the random quantity may be regarded as the statistically proper selection range [11]. Another example is the calculation of mathematical expectation equal to e for an infinite sequence of independent random values selected from a uniform distribution in the range $[0, 1]$ [12].

It comes that at $k > 2$ ($y < GRP_L$), the rate of energy exchange y is still too low to provide any ordering and organization of accumulated efficiency (information) Υ. The insignificant volume of accumulated information, in turn, does not support any reasonable form of its ordering and organization. It leads to the impossibility to use statistical prehistory, and *OTS* remains in a chaotic mode of development. In contrast to that, transition to the least k ($y > GRP_L$) signifies attainment of balance between accumulated Υ and achieved capabilities of *OTS* (rate y), which comes in the form of optimal energy decomposition in operation of *OTS*.

As in early agenesis ($k >> 1$), *OTS* is under strong influence of incoming y_{in}, then support of *OTS* internal energy integrity requires highly effective energy-saving mechanisms (high inheritance factor E^C) compared to supporting of sporadic changes (low variability factor E^T), i.e., $E^C >> E^T$, confirmed by direct estimation of (A.7, A.8). Afterwards, outcoming y_{out} gradually increases, finally achieving parity with y_{in} in the vicinity of *GRP*, which is in agreement with the state of dynamic balance discussed earlier. Arguing in this way, one is needed to recognize again that effective management of outcoming flow is one of the main signs of ordered exchange in *OTS*.

In accordance with the above, the presence of a pronounced boundary between genesis and agenesis suggests an idea of the existence of different ways of utilizing incoming energy on different stages of the energy spectrum. It is bearing in mind probabilistically more preferable collective regime in agenesis and individual one in genesis. Such an assertion gains even more sense if we agree that collectivity in organization of obtained energy is a forced measure and at agenesis is tied with low accuracy of copying between adjacent random spectral nodes at practically infinitesimal y-distance between them. Then it turns out that the collective form of organization for energy exchange is the only right decision for development at stage of the agenesis. On the other hand, higher accuracy of copying in genesis is naturally in line with finite y-distance between adjacent spectral nodes y_n.

Then, it would be logical to avow that another evolutionary purpose of Υ is functioning as cumulative or collective memory. This implies that collective memory "writes" and "saves" all previous states with both successful and failed evolutionary consequences. Then, every effort that *OTS* made assumes the accumulation of more Υ, which facilitates the fastening of deviations from the symmetry of probable outcomes, thereby pushing *OTS* to the new level of sophistication.

Of understandable interest is to compare the y-dependence of total energy exchange T (Figure 2.2 (a)) with efficiency Υ (Figure 2.2 (b)). It is seen that the stage with discrete spectrum in Υ ends much earlier (at LSP_R), then the positive stage for T (at TP). It occurs because of $\Upsilon(GRP_R) = 0$ in contrast to total energy T, which is zero at $TP = e^{3/2}$. It makes Υ harmonic in LSP_R the latest really visible, which confirms that discreteness of spectrum goes away as soon as Υ becomes negative irrelevant of changes in general exchange energy T.

It is logical to assume that the observed paired appearance of spectral nodes (2.17) may indicate the cyclic essence of energy decomposition in *OTS*. In truth, solution (1.12.a) in Figure 2.2 (b) clearly demonstrates non-monotonic character Υ as well as general T (Рис. 2.2 (a)), which confirms passage of solution over the same Υ and T points on different stages of *OTS*. Overall, it resembles change in evolutionary quantity during one full lifetime cycle. From the viewpoint of the functioning of *OTS*, it is self-evident that, as a whole, the dynamics of *OTS* is similar to the behavior of many living systems due to natural aging.

It is important to mention that from an optimization standpoint, condition (2.19) by default introduces the reference line $|\delta\Upsilon| = \int dU/Q = 1$. At first, this line means separation of the interval $[0, TP]$ (Figure 2.2) into three interleaving stages 1-3. Secondly, changes ~ $ln\ y$ in y-interval of discreteness are measured and evaluated in respect of this reference line. Obviously, that comparison of $|ln\ y|$ with 1 on stage 1 ($|ln\ y| >> 1$) and stage 3 ($|ln\ y| > 1$) does not make much sense, and this is one more reason to form the discrete spectrum in stage 2 only, where $|ln\ y| \sim 1$.

Bringing together the abovesaid, it is possible to notice one more important difference between, on the one hand, stages 1, 3, and, on the other hand, stage 2. It is the appearance of the non-zero volume of microstates in the phase space x-y that supports the droppable thermal energy. It looks like *OTS* tries to get rid of the useless energy in order to improve the structure of its energy exchange. Another argument in favor of this is the phenomenon of rotation of an eigenvector y during stage 2 that also indicates the reconfiguration of energy flows y along directions inwards and outwards *OTS*.

Also pay attention to the management of evolutionary scenarios by entropy mentioned in Chapter 1.2. It directly refers to the self-organizational capacity of *OTS*, which is quite important in evolution.

In the author's opinion, the features of *OTS* development discussed above give the reasons to think of *OTS* evolution as the process of going through a consequence of the predetermined mandatory steps of different importance.

In the above sense, how far (how many mandatory steps to do) *OTS* can move along the y-axis depends on the peculiar features of energy exchange and cannot be guaranteed at all. What, however, can be guaranteed is that there is no other way to move forward than going through the aforesaid mandatory steps. At this, it is not important how much time system dwells in each node or phase; what really does matter is that no evolutionary milestones can be missed.

Features of *OTS*, presented here, give grounds to think on the energy decomposition of *OTS* as a roadmap for realization of this or that evolutionary scenario.

In conclusion, it appears that the *CEL* can be used for the purpose of simulating the general course and features of the *OTS* evolution. Discovered features permit to define the energy infrastructure of evolution in the form of the set of spectral elements of the exact value such as nodes, phases, and stages. What is even more important, *CEL* enables to interpret the physical meaning of the most significant spectral elements.

Conclusion of Chapter 2.2

1. Nodes of the discrete energy spectrum with $n \leq 3$ possess pronounced bifurcation properties compared to the nodes with $n > 3$.
2. The energy spectrum of *OTS* consists of three basic stages. The second stage is composed of an infinite number of smaller phases, which altogether with the first and third stages permit speaking about the phenomenon of phasing of energy development in *OTS*.
3. The presence of predefined spectral elements (stages, phases, nodes, and internodal space) indicates the existence of infrastructure in energy development of *OTS*.
4. Model *CEL* describes a full lifetime cycle of *OTS*, from the stage of quasi-equiprobability $y \in [0, GRP_L]$ to the stage of expressed decay at $y > GRP_R$. In the range $[GRP_L, SP]$, entropy diminishes, while in the range $[SP, GRP_R]$, entropy grows.
5. In considering context, the ordering of chaos reappears as the emergence of the set of fixed in *y* bifurcation points along with predetermined connections between parameters of energy development.
6. In each *y*-point of evolution, energy *E* is a sum of two components, E^C and E^T. These components play a dissimilar role in evolution. Component E^C supplies an inheritance factor of energy exchange, while E^T does it for variability.

Chapter 2.3

Breaking of Time Symmetry as a Factor of *OTC* Energy Evolution

Below will be shown that the probabilistic background, which is necessary for the deployment of energy evolution, naturally stems from the employed model with an infinite number of conserved energy links with environment (1.2). This probabilistic background directly follows from the violation of time symmetry of the physical state *OTS* while preserving symmetry for the equation of motion. Besides, the breaking of symmetry can be used to logically explain both the fine structure of evolutionary energy and the evolutionary role of its components.

It should be noted that some plots in this chapter reproduce with some deviations those that are given in App. A. It is made deliberately to present in great detail the recent results through the lens of early relationships.

2.3.1. Decomposition of Evolutionary Energy

Based on (A.3), introduce the partitioning parameter for evolutionary energy (4.18) (parameter of fine structure)

$$\eta_n = \frac{E^C{}_{n\,min}}{E^T{}_n} \tag{6.1}$$

and rewrite (A.3) in the form

$$\eta_n(\eta_n + 2) - k_n{}^2 = 0 \tag{6.2}$$

with an obvious solution

$$\eta_{n_{1,2}} = -1 \pm \sqrt{1 + kn^2} \tag{6.3}$$

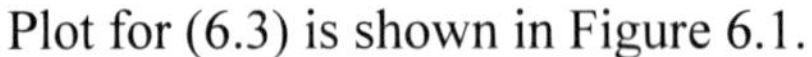

Plot for (6.3) is shown in Figure 6.1.

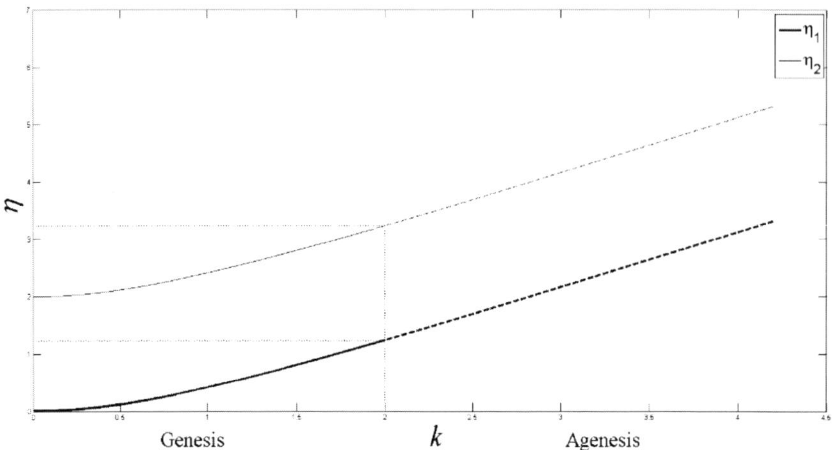

Figure 6.1. Dependence of factor $\eta_n = E^C{}_n/E^T{}_n$ on *k*. By an abscissa axis factor *k* (1.12.c) is indicated, by an ordinate axis factor η is indicated. Dash marks change at $k > 2$ when formally *k* is not defined. The form of curve $\eta(k)$ is in agreement with the assumption that in the course of evolution (diminishing *k*) energy E^C (inheritance factor) decreases while E^T (changeability factor) grows. In this sense, evolution is a process of finding an optimal ratio η.

Relation (6.3) reveals that in evolution, factor η_n experiences regular changes. At that, steady energy exchange modes appear only at some fixed ratio E^C: E^T, which points to predetermination for realization of the evolutionary case under the condition of a fixed *n*.

2.3.2. Infinite Splitting of Evolutionary Energy

As follows from (4.18.a), total exchange energy can be given as

$$T = E_o + E \tag{6.4}$$

where E_O is numerically equal to the area of asymmetric oval segment *ADBA*, *E* is numerically equal to the area of right-angled ΔACB. As of now, the details of the analytical representation of E_O are not essential; *E* is a dynamic combination of dropping and total energy. Study of dynamics and the structure of *E* will be the main purpose of this chapter (Figure 6.2).

According to (6.3), factors n and k are linked. It means that at fixed n (k), ratio η can have only two unique deterministic values (excepting case $k = 0$), as shown in Figure 6.3 by two bold-faced segments, D_nC_n and C_nF_n. As points *PoE* (introduced in section 1.2.1) coincide with the points of discrete k_n-spectrum (2.3), then due to (6.3), these points correspond to the nodes of discrete spectrum η_n.

Irrelevant of (A.3), area $\Delta D_nC_nF_n$ can be additionally split into an infinite number of the conjugated m-triangles (energies E_{nm}) which meet

$$E^C{}_{nm} + E^T{}_{nm} = E_n = const \tag{6.5}$$

where $m = 1, 2, \ldots$ at each fixed n. Random energies E_{nm} are outlined in Figure 6.3 by the conjugated $\Delta AC_nm_{i\ldots k}$ (for change $\Delta E_n < 0$) and $\Delta AC_nm_{w\ldots z}$ (for change $\Delta E_n > 0$). Also, each E_{nm} can be split into an appropriate pair of $E^C{}_{nm}$ and $E^T{}_{nm}$ in conformity with the meaning of random η_{nm}.

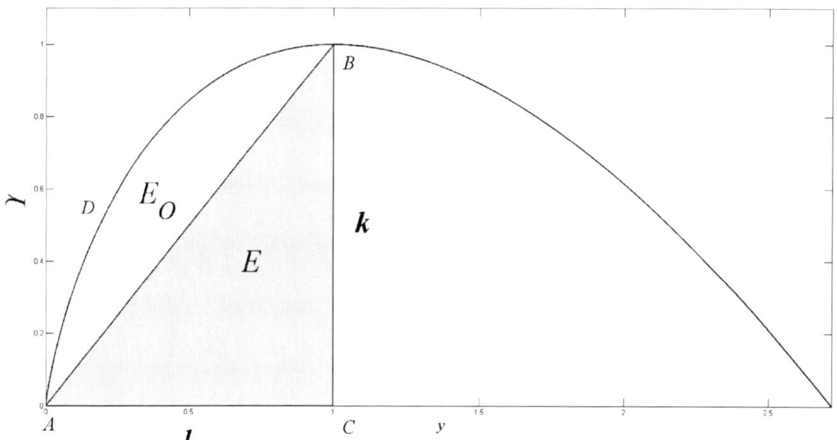

Figure 6.2. Total exchange energy T as sum of energy E (triangle ACB marked in dark grey) and E_O (oval segment $ADBA$ marked in light grey). In the plot, by an abscissa axis an energy exchange rate y is indicated, and by an ordinate axis the integral efficiency of energy exchange Υ is indicated.

It is important to know that η_{nm} is strictly within the interval $[\eta_1, \eta_2]$ for each *PoE*. In the non-*PoE*, the continuous filling-in of the range (η_1, η_2) occurs.

Now we can combine n- (6.3) and m- (6.5) splits into one set populated with random $E^C{}_{nm}$ and $E^T{}_{nm}$ (further for clarity also called $\hat{E}^C{}_n$ and $\hat{E}^T{}_n$) within the limited range $[\eta_1, \eta_2]$. So, (6.3) defines the n-range for all possible m changes.

Physically, we can think of energies $\hat{E}^C{}_{nm}$ and $\hat{E}^T{}_{nm}$ as the degenerate set of the eigenstates E_n, which all share the same energy E_n.

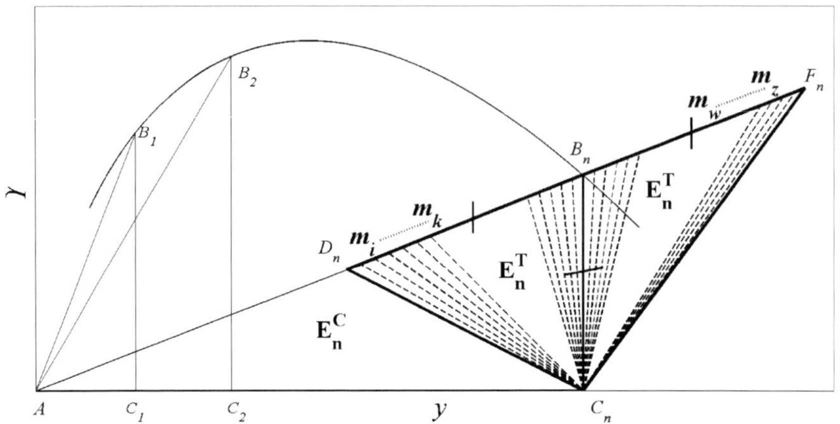

Figure 6.3. Curve $\Upsilon(y)$ with ΔAF_nC_n and an infinite family of conjugated m-triangles at fixed n. Each ΔAF_nC_n of area $E_n^C+2E_n^T$ can be split into an infinite number of conjugated m-triangles within the range $[\eta_1, \eta_2]$. $\Delta D_nC_nF_n$ shown in bold contains all possible positions for random η_{nm}. Marginal segments C_nD_n and C_nF_n show position of solution (6.3). Intermediate segments C_nm_r show position of solution (6.5).

In its connotation, $E^C{}_{nm}$ and $E^T{}_{nm}$ are close to casual understanding of spontaneous symmetry breaking (*SSB*) when the result of physical action is determined by the result of $E^C{}_{nm}$ and $E^T{}_{nm}$ shuffling to optimize ratio η and make $\eta = \eta_n$ in agreement with (6.3).

At this, the degeneracy (6.5) comes as a consequence of continuous symmetry, and we can believe that the energy rigidity of the system is governed by the set (6.5).

Below, to formalize handling of the random quantities, we stick to a uniform probability distribution for them.

The question to come is about the analytic form for E_{nm} transformation. The answer to this question is discussed in the next section.

2.3.3. Energy Form of Tangent-Secant Theorem

As follows from the section A.4, the support ranges for random $\hat{E}^C{}_n$ and $\hat{E}^T{}_n$ overlap. Therefore, the sums $\hat{E}^C{}_n$ and $\hat{E}^T{}_n$ can be given as the algebraic sum of deterministic (at fixed n) $E^C{}_{n\,min}$ $(E^C{}_{n\,max})$ and the random $\hat{E}^T{}_n$, so that

$$\begin{cases} \hat{E}_n^C + \hat{E}_n^T = E^C{}_{n\min} + \hat{E}_n^T \\ \hat{E}_n^C + \hat{E}_n^T = E^C{}_{n\max} - 2\hat{E}_n^T \end{cases} \tag{6.6}$$

where $E^C{}_{n\,max} = E^C{}_{n\,min} + 2E^T = E_n + E^T$, and $\hat{E}^T{}_n$ on the right-side (6.6) changes within the joined range of random $\hat{E}^C{}_n$ and $\hat{E}^T{}_n$, which is $[0, E^T{}_n]$.

Then, using (6.5), (6.6), we obtain the basic equation for stochastic energy exchange (*SEE*)

$$E^C{}_{n\min}(E^C{}_{n\min} + 2\hat{E}_n^T) = (k_n\hat{E}_n^T)^2 \tag{6.7.a}$$

$$E^C{}_{n\min}(E^C{}_{n\max} - 2\hat{E}_n^T) = (k_n\hat{E}_n^T)^2 \tag{6.7.b}$$

as shown in Figure 6.4.

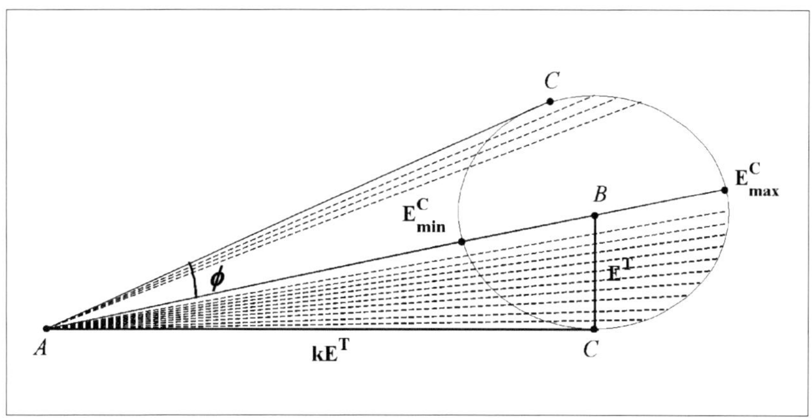

Figure 6.4. Tangent-secant form of ratios (6.7.a,b) at $\delta_{ik} = 1$. An infinite number of random *m*-secants for fixed *n* is shown by dashed lines. $E^C{}_{min}$ and $E^C{}_{max}$ mark the range of random variations for $\hat{E}^C{}_n$ and $\hat{E}^T{}_n$. The deterministic secant $\delta_{ik} = 0$ is shown by the solid line through center *B*.

Now, associate area $E^C{}_{n\,min}$ with the external part of the secant, sum $E^C{}_{n\,min} + 2\hat{E}^T{}_n$ or $E^C{}_{n\,max} - 2\hat{E}^T{}_n$ with the whole secant, $k_n\hat{E}^T{}_n$ with the tangent, then (6.7.a,b) is identical in its form to the well-known tangent-secant theorem as shown in Figure 6.4. In Figure 6.4, the random secants are shown by the dashed lines and the deterministic secant through the centre of circle B by the solid line.

The significance of (6.7.a,b) is that it is the combined mathematical form of (A.3) and (6.5), which provides the formalism for E_{nm} combined n- (A.3) and m- (6.5) transformation.

Further, we will be working with the form (6.7.a) only, although all results remain equally applicable for the form (6.7.b) either.

Also, pay attention that (6.6, 6.7) work for the area only, not for the triangle's legs. In other words, (7.7, 7.8) are not valid for linear quantities such as y and Υ, which indicates the fundamentally nonlinear essence of (7.7, 7.8).

Now, it is worth noting that the length of legs in the ΔABC is numerically equal to the length of appropriate legs of the K_n-triangle multiplied by E^T (Figure 6.4). So, $\Delta ABC \propto K_n$-triangle mentioned in the section 2.1.2. Hence, further simplification of calculation we will be considering ΔK_n in the belief that all conclusions are applicable for ΔABC either. To underline the new meaning of ΔK_n, further its elements will be displayed in bold.

2.3.4. Evolutionary Meaning of Tangent-Secant Theorem

Based on (6.7.a,b), consider the meaning of the following quantity presented in the complex form

$$R = E^C{}_{n\min} + 2\hat{E}_n^T = E_n + E_n{}^T exp\left(i\delta_{kk_n}\hat{\varphi} \right) \tag{6.8}$$

where random $\varphi = 2arctan(1/k)$, Kronecker delta

$$\delta_{kk_n} = \begin{Bmatrix} 0 & if\, k = k_n \\ 1 & if\, k \neq k_n \end{Bmatrix} \tag{6.9}$$

shown in Figure 6.5.

In a physical sense, (6.8) describes a random change of energy E_n in the range $[E_n - E^T{}_n, E_n + E^T{}_n]$.

Accounting meaning of $E^T{}_n$, we should think that quantity R manages all allowable variations of E_n. Then, staying within the considered paradigm, we can believe that R is the random value of energy on the next step of evolution $n+1$, i.e., finally, rewriting (6.8)

$$E_{n+1} = E_n + E_n{}^T exp\left(i\delta_{kk_n} \hat{\varphi} \right) \tag{6.10}$$

So, (6.8 – 6.10) can be thought of as an equations of energy evolution for considered *OTS*.

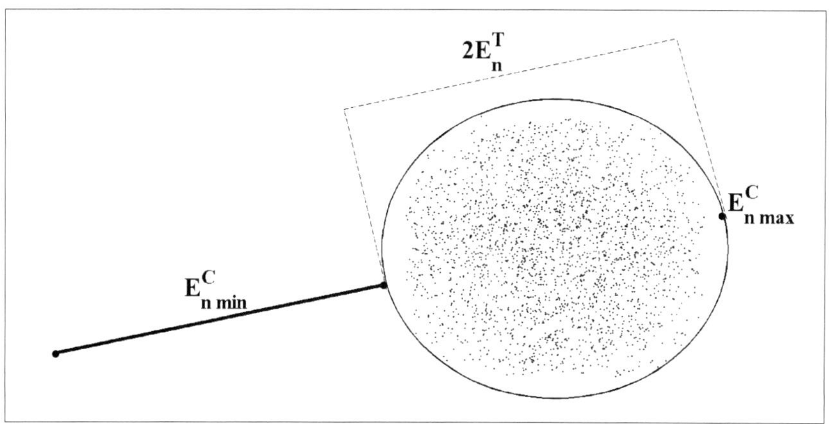

Figure 6.5. The same as in Figure 6.4 presented in the complex form. The range for random $\hat{E}^T$ and $\hat{E}^C$ is shown as a circle with diameter $2E^T$. Minimal and maximum possible energy E_{n+1} at given E_n are also shown.

2.3.5. Emergence of Breaking for Time Translation Symmetry Breaking (*TTSB*)

In [13], it was paid attention to the specific non-thermal nature of uncertainty of (1.2) at $n \rightarrow \infty$. It was suggested to consider discovered uncertainty in the context of the energy exchange process based on the essentially non-thermal fluctuations. In this sense, discovered uncertainty coexists with the symmetrical thermal uncertainty.

In accordance with all mentioned above for the link between energy conservation law and symmetry of temporal translation, we are almost obligated to think of (6.5-6.8, 6.10) through the prism of time translation symmetry breaking (*TTSB*). We demonstrate it below.

Fix cross section A for energy flux y, then, ultimately, y change is determined by the product $t{\cdot}E$ as per (1.1). Then for fixed y, the change in E causes an appropriate change in t. According to (6.10), at $k = k_n$, energy

$$E_{n+1} = E_n + E_n^T \tag{6.11.a}$$

is the deterministic single-valued quantity. Geometrically, it fits to *PoE* when all random secants precisely collapse to the only secant through the centre of circle B (Figure 6.4). That is why in *PoE* we do not see any signs of *TTSB*.

On the contrary, at $k \neq k_n$, energy

$$E_{n+1} = E_n + E_n^T exp\left(i\hat{\varphi}\right) \tag{6.11.b}$$

is the random quantity in the range $[E^C{}_{n\ min},\ E^C{}_{n\ min}+2E_n{}^T]$.

Geometrically, it means that there are multiple random secants in circle B, and we observe the multitude of possible $\hat{E}^C{}_n$ and $\hat{E}^T{}_n$ combinations shown in Figure 6.4.

In other words, in *PoE* we have a perfect balance between bidirectional random energy fluxes y_n incoming and outcoming *PoE*, so there is no need to do anything else to meet a cohesion of energy fluxes. That is why the discrete implementation of the factor k is in use (2.5). However, in the non-*PoE*, to fix the mismatch between random y_n and keep (1.6.a), the system ought to generate additional time-energy paths t_{nm}-E_{nm}, which cannot be realized by anything but breaking of energy and temporal symmetry (Figure 6.6). That is why the continuous implementation of the factor k is in use (1.12.c).

Based on the above, we can expect the appearance of the "time crystals" in *PoE*, which defines the position for the moving frame of reference. Assuming the position of *PoE* at (t->t+nT{\displaystyle t\to t+nT}), we can introduce the driving period between the "time crystals" in the logarithmically equispaced y-grid as

$$T(n) = PoE_n - PoE_{n+1} = exp(\pm\frac{1}{n^2+n}) \tag{6.12}$$

So, the physical pattern will be quasi-periodic with the driving period (6.12) as schematically shown in Figure 6.6.

Accommodating recent changes, aggregate energy (6.4) can be given as

$$T = E_O + E_n + E_n^T exp\left(i\delta_{kk_n}\hat{\varphi}\right) \tag{6.13}$$

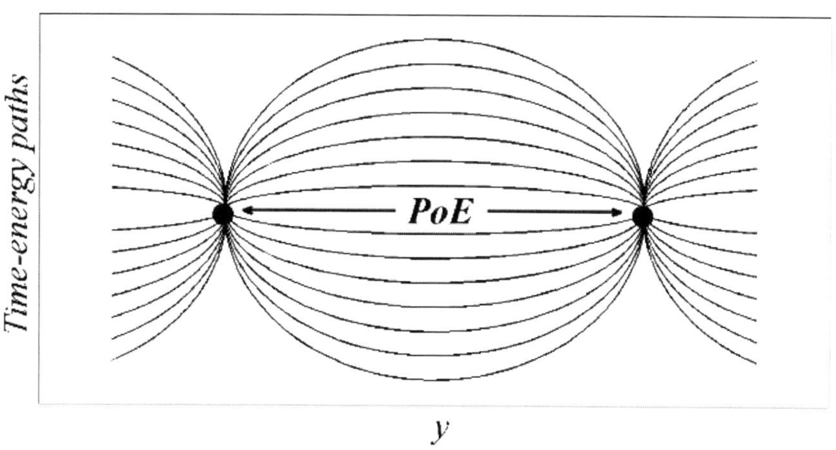

Figure 6.6. Diagram for multiple time-energy paths t_{nm}-E_{nm} between *PoE*. Diagram is built on the assumption that $y_{nm} \sim t_{nm} \cdot E_{nm} = const.$

2.3.6. Partitioning of Evolutionary Energy

In this section, the arguments in favor of the asymmetrical role of components E^C (A.7) and E^T (A.8) in the evolutionary process are considered.

1. To confirm the set-up hypothesis on the asymmetrical role of components E^C and E^T, we can start from (A.11) in the section (A.5). Rephrasing it in terms of energy, we may state that the range for random changes of $\hat{E}^C$ is all the time less than that of $\hat{E}^T$.
2. Take a look at dependence for a swing of fluctuations $\hat{E}^C$ and $\hat{E}^T$ on k (Figure 6.7). It is seen that in the course of y, component $\hat{E}^T$ demonstrates steady excess swing compared to $\hat{E}^C$.

From this angle, look at the evolution equation (6.10). It shows that the *SEE* pattern in *OTS* translates to the next n with the rescaled $\eta = E^T{}_{n:}\ E^C{}_n$ but keeping the same functionality due to (6.1), i.e., evolution demonstrates intrinsic inheritance behavior. At the same time, the new value of E_n is a principally random quantity, i.e., evolution shows intrinsic variation behavior either.

3. Consider the *SEE* pattern in detail. With this purpose, compare variance D for E^T_n and E^C_n. Remind that $E^T_n \in [0, 2E^T_n]$ while $E^C_n \in [E^C_n, k_nE^T_n]$. Then, for the declared uniform distribution [21] in k-basis

$$D_{E_n^T} = \frac{1}{12}(2E_n^T)^2 \tag{6.14.a}$$

$$D_{E_n^C} = \frac{1}{12}(k_nE_n^T - E_n^C)^2 \tag{6.14.b}$$

Then an inequality

$$D_{E_n^T} > D_{E_n^C}$$

is equivalent to

$$\sqrt{k_n^2 + 1} + 1 \geq k_n$$

which holds at any k_n.

Therefore, in the pair $\{E^C_n, E^T_n\}$ energy E^T_n is essentially more variable compared to E^C_n, which confirms the difference in evolutionary dynamics of $\hat{E}^T$ and $\hat{E}^C$.

4. Finally, compare the physical meaning of E^C_n and E^T_n as stems from above. With this purpose, turn to η-basis using (6.1). Then, quantity E^T_n determines how much energy can be invested in or taken out of *OTS* to support the next step of evolution and, which is also important, not destroy established energy structure. In order to meet these conditions, E^T_n should respect $E^C_{n\,min}$ and $E^C_{n\,max}$, which delimits the range of m-fluctuations, i.e., E^T_n changes within the borders created by E^C_n. Logically, E^T_n can be zero while E^C_n cannot ever be zero as

the statement $E^C{}_n = 0$ is equivalent to the physical absence of the even minimally stable *OTS*.

Hence, $E^C{}_{nm}$ appeals to the energy exchange part that supports the conservative (inheritance) properties of *OTS*, while $E^T{}_{nm}$ does it to provide the permanent and consistent changeability. It suggests that $E^T{}_n$ compared to $E^C{}_n$ is more unstable, which is close to the key difference, for example, between inheritance and variability in natural selection [24].

In the light of the above, it is easy to understand the above statement that if we keep the same E_n, then the result of applying the pair high inheritance-low variability energy (high η) can be completely different from applying the other m-pair with strong variability-weak inheritance (low η).

Also, it is clear why, within the considering approach, the result of the same E_n action crucially depends on whether $E^C{}_{n\,max} > E_{nm} > E^C{}_{n\,min}$ or otherwise. If it is, n remains the same. If not, n shall change to $n - 1$ or $n+1$ depending on which boundary, $E^C{}_{n\,min}$ or $E^C{}_{n\,max}$, random E_{nm} hits. This is the mechanism to protect the course and existing structure of evolution. So, the result of evolutionary action produced by E_n clearly depends on m.

Despite noted differences, as a whole, in the evolutionary process, the pair $\hat{E}^C$ and $\hat{E}^T$ work together, providing the evolutionary needs of *OTS*.

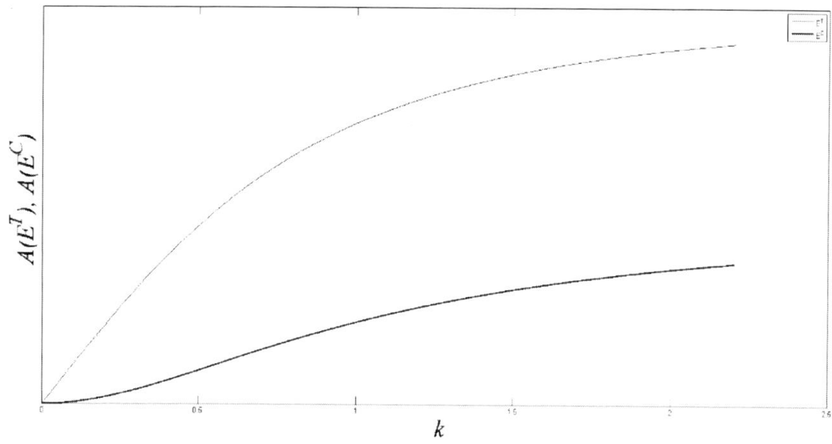

Figure 6.7. Dependence of a swing for fluctuations $\hat{E}^C$ and $\hat{E}^T$ on factor k. In the plot, by an abscissa axis factor k is indicated, and by an ordinate axis a swing of energy fluctuations $A(\hat{E}^C)$ and $A(\hat{E}^T)$ is indicated. It is seen that inequality $A(\hat{E}^T) > A(\hat{E}^C)$ retains at any $k > 0$. Profile for $A(\hat{E}^C)$ shown in bold font, while for $A(\hat{E}^T)$ in regular font.

The above-listed content supports the hypothesis set up in the beginning of the current section that energy components $\hat{E}^T$ and $\hat{E}^C$ play an unlike role in the evolution of *OTS*. More generally, $\hat{E}^C$ provides continuity and inheritance of energy structure, while $\hat{E}^T$ supports its changeability.

2.3.7. Role of Time Symmetry Breaking in Evolution

From the standpoint of (6.3), it is possible to assert that the energy evolution of *OTS* manifests as a combination of random energy modes at $y \neq y_n$ and deterministic modes at $y = y_n$. In its turn, the presence of the sequence of ordered energy modes makes it possible to discuss the existence of a relatively stable carcass or infrastructure of energy evolution, as it was considered in detail in chapter 2.2. Of course, the availability of each node in this infrastructure is determined by the achieved level of energy exchange rate in *OTS* and, generally, cannot be guaranteed.

Pay more attention to ratio (6.7), and with this purpose, turn to Figure 6.8. The theorem in section A.7 reads that the maximum ratio of the internal part of the secant to the total length of the secant is observed if the secant goes through the center of the circle. Also, if the secant crosses the center of the circle, the deterministic mode y_n of evolution is observed. All the other secants (for example, BB_1 and BB_2) corresponding to random solutions just create the necessary probabilistic background. In translation to energy language, it is equivalent to the scenario when, for a given n, the deterministic modes have the maximum possible (diameter of circle) the range of random changes (thereby supporting the variability of energy E^T). In the opposite case, the more predetermination in value and sign of energy mode exists, the less evolutionary potential this mode has and the fewer chances it has to survive.

In this sense, there exists an understandable connection between a particular method of partitioning of evolutionary energy (fine structure of spectral mode) and its evolutionary stability.

Return to the meaning of ratio (A.2), which declares dynamic balance between $2E^T$ and E^C at $y = GRP$. Now, it presumes a new interpretation showing that a dynamic balance between stability (E^C) and variability (E^T) acts as a necessary criterion for the really qualitative leap in *OTS* development, supporting the optimality of evolutionary modifications.

In a similar way, (5.5) can be interpreted as an indication that dynamic balance in the fine structure of the exchange process between the dropping energy and total energy at y = *GRE* is critical for the qualitative leap in evolutionary functionality *OTS*.

At the bottom of the fact, the equations of energy evolution (6.8-6.10) are just modified (A.3) that was extended over the domain of random energies. In this respect, the critical element of our reasonings in favor of the evolutionary nature of (6.8-6.10) is the availability of *mn*-evolutionary infrastructure (6.3, 6.5).

This approach demonstrates how the geometry of the problem works jointly with Hamiltonian physics, and evidence for the credibility of this approach is confirmation of the possibility of energy evolution in the spontaneous development of *OTS*.

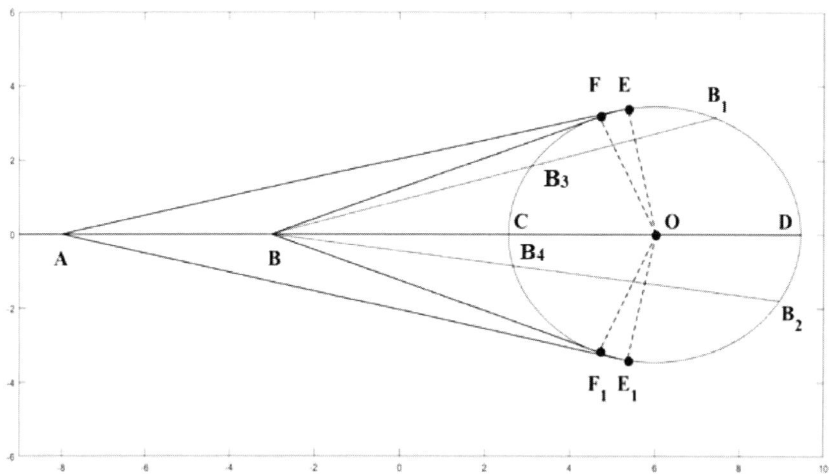

Figure 6.8. Geometry interpretation of equation (6.7) for energy evolution. Evolutionary triangles E_1AO, EAO, F_1BO, and FBO for consecutive evolutionary steps are shown; point O is the centre of the circle. The tangents AE and AE_1, AF and AF_1 correspond to kE^T, the secant through the centre of the circle to $E^C + 2E^T$. Diameter CD is numerically equal to the range $2E^T$. The secants through point O correspond to the most stable discrete modes of the energy spectrum with $y = y_n$, whereas other secants, for example, BB_1 and BB_2 (dashed lines), fit to less stable modes $y \neq y_n$, which are being suppressed at the forming of the spectrum. The minimal possible value $S^C{}_{0min}$ at a given k is shown by segments BC and AC.

In that sense, (6.10) creates proper probabilistic background, where physical phenomena such as evolution could be understood, realized, and formalized. At that, a probabilistic background here can be considered a necessary condition for the existence of a driver in energy evolution *OTS*.

In this connotation, to assert that geometry provisions were used only for convenient interpreting of physical results would not be accurate. Of course, studied *PoE* are the points of bifurcation (*PoB*) (2.5), but at the same time and independently, *PoE* are the points of pronounced geometrical effects [22]. Hence, the existence of the notable *PoE* is a fundamentally interdisciplinary result. In this sense, the points (2.3, 2.5) clearly separate *PoE* from the other points.

On the one hand, in *PoE*, we see the difference between the time translation invariancy for the equations of motion (1.1) and (A.3) and, on the other hand, the non-invariancy in the actual physical state, which leads to the appearance of the time-energy asymmetry t_{nm}-E_{nm} (6.5). This fact is quite common for the phenomenon of *SSB* [15].

It is worth noting that *TTSB* comes as an uncertainty in the space of two dimensions and allows easy extension in the space of three dimensions. Removal of the second dimension symmetrizes applied transformation and prevents the emergence of an evolutionary potential. Therefore, to see the full results of the presented model, we should guarantee the two-dimensionality of the model, at least. So, the concept of two-dimensional area-energy is essential for this approach.

Another aspect of the discovered *TTSB* is that it occurs in the non-equilibrium medium far away from thermal equilibrium at all the points (excepting the point of stationarity *SP* at $y = 1$). This perfectly matches existing understanding of which conditions should be met to observe *TTSB* [25].

It is also worth noting that in terms of *m*-uncertainty, the result of *SSB* transformation is principally different at different *m*

$$E^{C}{}_{nm1} + E^{T}{}_{nm1} \; \{\neq\} \; E^{C}{}_{nm_2} + E^{T}{}_{nm2} \tag{6.22}$$

where the symbol $\{\neq\}$ denotes the non-equivalence of the physical results, $m_1 \neq m_2$. In other words, inequality (6.22) means that at fixed *n*, the result of evolutionary action produced by E_n depends on $m(\eta)$.

In this connotation, the meanings of $E^{C}{}_{nm}$ and $E^{T}{}_{nm}$ are close to convenient understanding *SSB*, when the result of physical action is the result of a reshuffle of terms $E^{C}{}_{nm}$ and $E^{T}{}_{nm}$ in order to optimize η and create favorable conditions for $\eta = \eta_n$ (6.3). At that, the degenerate values (6.5) act as a consequence of the original $[0, GRP_L]$ continuous symmetry and provide the required energy stiffness for *OTS* on the stage of the discrete spectrum *[GRP_L, GRE_R]*.

In the above sense, accounting for the results [17], (6.10) can be interpreted in the way that E^C_{nm} and E^T_{nm} compose the non-commutative group. The importance of the relation between the group non-commutativity and the phenomenon of *SSB* is the well-known fact [18].

Conclusion of Chapter 2.3

1. The existence of m-splitting logically supplements and completes the n-infrastructure of energy exchange *OTS* discussed in Chapters 1.2 and 2.2.
2. There exists a fundamental relation between provisions of Euclidean geometry, Hamiltonian mechanics, and non-equilibrium thermodynamics in the paradigm of the energy conservation law, which, in part, manifests in the phenomenon of energy evolution through the mechanism of *TTSB*. At that, *PoE (PoB)* of Hamiltonian are the points of equilibrium (static and dynamic) and, at the same time, the points for delimitation of the evolutionary process in *OTS*, in which the "time crystals" determine the position of moving reference frames.
3. Evolutionary energy E is divided into two parts, which play dissimilar roles in evolution. The existence of the fine structure in evolutionary energy creates prerequisites for the emergence of optimization m-uncertainty, which establishes the basis for competition between factors of inheritance and variability in the evolutionary process. Such competition supplies a driving force of evolution in the stage of discrete spectrum.
4. In the presented theory, *TTSB* is revealed as a forerunner of energy evolution as it creates the probabilistic background that enables the emergence of the driving force for further energy development.
5. In *PoE (PoB)*, we observe the exact balance between bidirectional random energy flows y_n incoming and outcoming *PoE*, therefore, there is no necessity to make any additional steps to meet a cohesion of energy flows. However, in the non-*PoE*, to fix the mismatch between random y_n and keep (1.6.a), one ought to generate additional time-energy paths t_{nm}-E_{nm} which cannot be realized by anything but breaking energy and temporal symmetry.

6. For continuous evolution, an existence of the non-monotonous dependence $k_n(n)$ is necessary. Otherwise, though identity k_n provides ideal protection of the profile in the transferred energy form and supports inheritance, it completely nullifies variability and excludes any chance for the appearance of evolutionary changes.
7. The principal possibility of evolution is realized within geometry defined by fundamental relation (A.3).

Part 3. The Physical Basics of Genetic Code and *DNA* in *CEL*

For obvious reasons, cells and nucleus (further cells) need energy to support their vital functions. Take a look at the known facts in the general pattern of energy exchange between cells and milieu.

Firstly, the energy exchange process is provided by the external and internal energy supplies/sinks of different essences, starting from quite generic sunlight and organic food, continuing with intermediate energy-rich molecules like *ATP*, and completing with more specialized photosynthesis, glycolysis, citric acid cycle, oxidative phosphorylation, and others [1].

Secondly, the response of cells in an energy exchange possesses some time lag and does not process the energy requests as soon as they come [2]. In reality, cells release and receive energy through a continuous series of reactions contributing to the creation of a persistent energy background.

Thirdly, some central energy-processed mechanisms, such as the citric acid cycle, work as a closed loop, i.e., the last part of the pathway reforms the molecule used in the first step [3]. It makes such mechanisms the continuously working machine providing a quasi-continuous flow of energy interactions.

Another important thing is that existing bidirectional [1-3] energy pathways manifest actual evolutionary coexistence. It is known that though glycolysis *(GL)* arose very early in evolution, it is still quite compatible and consistent with photosynthesis (*PH*) and oxidative metabolism (*OM*), and all present-day cells are to support *GL*. Release of oxygen as a consequence of *PH*, at the same time, is a precondition for the development of *OM*. At this, some researchers believe that *OM* may have evolved before *PH* due to the activity of ancient microbes, which means that oxygen was available for living entities in the epoch of *GL* dominance [4-6]. Hence, the mentioned energy pathways can work in accordance and can replace each other if required. The latter indicates that the boundaries between different energy pathways can be considered as floating and sufficiently conditional.

The point here is that energy exchange actually consists of a succession of many bidirectional random processes of smaller scope. Summarizing, an energy exchange between cells and milieu meets requirements for applicability of *CEL* theory highlighted in Chapter 1.1. So, below *CEL* will be applied for a description of the possible energy structure of genetic code (*GC*) and deoxyribonucleic acid (*DNA*).

Chapter 3.1

The Physical Constraints in the Origin of Genetic Code

3.1.1. Selection of Nucleotides

As it is known, genetic code (further *GC*) is not a random phenomenon [7], therefore, there exist some natural reasons for what we observe in the appearance of *GC* these days. Below, we focus on those consequences of the process of *SEE*, which look linked to *GC*.

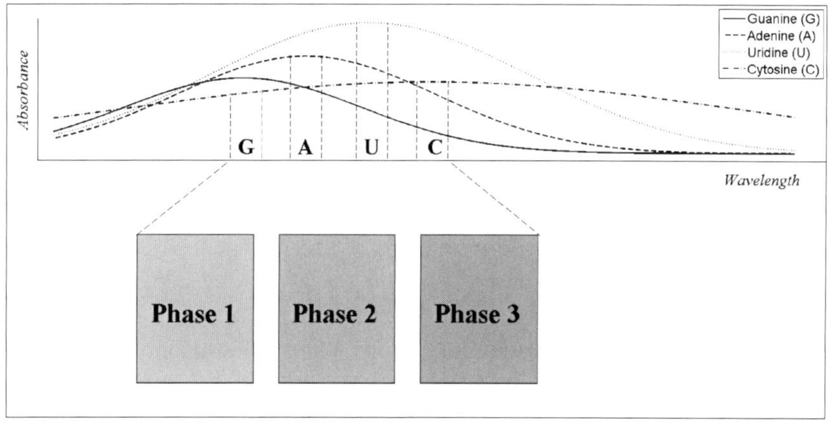

Figure 7.1. Schematic *UV* spectra for absorbance in adenine, cytosine, guanine, and uridine nucleotides in dependence on wavelength. 10% of the areas around maximum absorbance are marked with vertical dashed lines. Peaks of absorbance fit to guanosine phosphates (marked *G*), adenosine phosphates (marked *A*), uridine phosphates (marked *U*), and cytosine phosphates (marked *C*).

By its origin, nucleotides are organic molecules composed of a nitrogenous base, a pentose sugar, and a phosphate. They serve as monomeric units of the nucleic acid polymers *DNA* and *RNA*, both of which are essential biomolecules within all life forms on Earth [8]. The *RNA* world hypothesis is a currently predominant scientific platform for analysis of the early evolution of biological and pre-biological structures, the main advantage of which is the assumption that *RNAs,* as the first living systems, were self-sufficient as both catalysts and templates [9]. It has been argued that an increment of three for the number of nucleotides is a fundamental physical property of *RNA*, so that the translation system originated as a triplet-based mechanism [10, 11]. According to the *RNA* world hypothesis, free-floating ribonucleotides were present in as early as the primordial soup. To be exact, the nucleotides *A, G, C,* and *T* are found in *DNA,* while *A, G, C,* and *U* are found in *RNA*.

If to account that higher absorbance is accompanied by easier breaking of hydrogen bonds and the formation of free nucleotides, then the best candidates for the generation of free nucleotides are the chemicals (for example, phosphates) with a high level of absorbance rate based on the adenine *(A),* cytosine (*C*), guanine (*G*), thymine (*T*), or uridine (*U*) nucleotide set (Figure 7.1). These five nucleotides compose the limited *GC* alphabet and have the privilege of being used as building blocks for nucleic acids and further *RNA* and *DNA*. Besides, cited nucleotides have relatively close wavelengths and high absorbance peaks, which altogether contribute to the emergence of them as a compact group. In other words, these five prime nucleotides become a convenient material to experiment with, in part, with different realization scenarios (section 3.1.8).

Of course, any manipulations with nucleotides, such as interconnection, assume the presence of some energy source. In the case of nucleotides, it is not a problem, as needed energy may be taken from the nucleotides themselves, which have three phosphates attached to each of them (much like the energy-carrying molecule *ATP*) as the energy sources. When the bond between phosphates is broken, the released energy is used to form a bond between the incoming nucleotide and the growing *DNA* chain. Thereby, the energy to drive the attachment of a nucleotide to the end of a growing *DNA* comes from phosphate-group transfers supported by *ATP*. Free-floating nucleotides are attracted to the exposed bases and bond to them via complementary base pairing (adenine-thymine and guanine-cytosine). *DNA* polymerase then joins the new nucleotides together in a series of condensation reactions, forming phosphodiester bonds in the sugar-phosphate backbone.

If the deoxyribonucleotides (*dATP*, *dCTP*, *dGTP*, and *TTP*) do not have their own energy, the molecule *ATP* needs to be split in order to link these deoxyribonucleotides together and ultimately form *DNA*. This process supplies necessary energy to the polymerization process in *DNA* replication.

So, from an energy standpoint, five noted nucleotides are capable of forming an energy pool of *4 (5)* available prime nucleotides. On one hand, they are clearly distinct from the others in the wavelength, on the other hand, their absorbance-wavelength (energy) signatures are sufficiently close to thinking of them as a group of independent constituents.

3.1.2. Genetic Context of Energy Evolution

The origin of the genetic system, in general, and *GC*, in part, is probably the central and most difficult problem in the investigations on the origin of life and one of the most complex problems in evolutionary biology as a whole [12, 13]. The evolutionary forces that produced the canonical genetic code before the last universal ancestor remain obscure [11, 14]. There are multiple hypotheses on the emergence and development of existing *GC* and related replication and translation, although none of them provides a description without gaps and assumptions [7, 9-16].

At the same time, it is an indisputable fact that, in order to comprehend how such a magnificent natural phenomenon as *GC* came into existence, an insight into physical pathways in addition to all others is extremely desirable [15]. Recognition of this obvious conclusion is one of the trends in modern evolutionary science.

It is well agreed that the essence of *GC* is to instruct the cell how to build proteins. Then, it is logical to believe that at least some form of *GC* was created and naturally integrated into the primordial environment prior to or simultaneously with the arrival of the proteins.

The opposite viewpoints, mainly, associate the appearance of *GC* with the coming of the building material for its working, i.e., the amino acids [16], for example, the theory of a “frozen accident” [13]. In any case, there are serious arguments to state that the mechanism for forming the original *GC* started as early as in abiogenesis and has become fully operational in biogenesis [15].

So, the main hypotheses for the origin of *GC* are (1) the error minimization theory, considering natural adjustment as a tool for evaluating the impact of mutation and translation; (2) the stereochemical theory, based on the physicochemical affinity between amino acids and anticodons; and (3) the coevolution theory, which states that the code structure coevolved with the biosynthesis of amino acids [16].

Despite their substantial contribution, the above theories do not provide sufficient arguments in favour of the existing numeral basis of *GC* exactly as we know it, i.e., *3, 4, 20,* and *64* [17]. In the meantime, going along with the physical way of seeing things, we are obligated to accept that these numbers reflect some fundamental property of nature.

From the above, it looks reasonable to consider the appearance of *GC* as an integral part of the general long-term evolution of nature, with the driving force of *GC* development based on the as ubiquitous and permanent elements of the Earth environment throughout the entire natural history as possible. Then, theories on the early phases of the evolution of *GC* should be constrained by the minimal complexity that is required of a self-replicating system, e.g., [18].

It should be noted that in presenting this chapter, we try to refrain from any comments on the chemical and biochemical factors that may or may not influence the discovered energy transformations due to the announced agenda of this book, which is mostly about the physical aspects of stochastic energy exchange (*SSE*) manifesting in *GC* and *DNA*. Of course, all the missing points are of the greatest interest, but it would also be interesting if some high-level theoretical background, usually provided by physics, could be involved.

Below, we rely on provisions of *CEL* theory mentioned in the introductory chapter 1.1, including all suitable energy transformations related to nucleotide-related processes.

So, the major purpose of this chapter is to review the physicomathematical grounds that, by working together, could favour the formation of a discrete numerical basis for known and some hypothetical former (future) numerical structures of *GC* that share the same physical paradigm. There are also some reasons discussed later to believe that considering physical concepts can be operative in the formation of an energy carcass [19] of *DNA*, as introduced in 3.2.3.

3.1.3. Nonuniformity of Energy Phase Space

In the course of this research, we stay in the early-declared position that very universal and simple mechanisms naturally integrated into the structure of the existing physical world should be recruited for accurate simulation of conditions on early Earth. That is why, below, we try to present possible meaning of the requirement of the minimal complexity mentioned earlier.

So, it looks like, playing with different physical mechanisms, nature was ultimately capable of inventing means to defeat ubiquitous chaos and maintain the uniqueness of prebiotical entities through the use of a limited discrete set of highly protected units. This means, partly considered in Chapter 2.3, is to implant protection of sensitive inheritance information directly into the structure of *SEE* between *OTS* and the environment, as is discussed later.

The important moment here is that the details of the early environment in which life on Earth originated are poorly known [20]. We do not know for sure what and how exactly it might happen in this young and brutal world. We may just imagine the severe milieu with mineral surfaces covered by some bituminous material exposed to hard *UV* radiation through a reducing atmosphere. That is why the unprejudiced consideration of evolutionary mechanisms should not depend on the assumption that substances and mechanisms, seemingly essential now, were essential initially. Equally, we have no idea about the actual energy-supplying mechanisms the primordial conditionally living units used. In this case, the working idea may be to replace the absence of actual detailed information with the utmost simplification, unless we face what we know for sure. Then, going this way, the sooner or later we press into the physical sentinels of our world, the physical laws.

Then, in this sufficiently simple environment unambiguously governed by physical laws, we may expect the dominance of relatively simple decision logic either. If all that primitive inhabitants were interested in was the existence or non-existence of some result, then, with all necessary assumptions and clauses, the yes-no logical copy of the existing world may be acceptable.

On the other hand, in its activities, natural evolution teaches us to make cautious use of what was designed and implemented before. That is why a life, and especially the prebiotic life, tried to stick to the mechanisms that have been developed, testified to, and proved, some more, some less, their evolutionary value. In part, it may mean that nature prefers to use rescaling, i.e., compressing and decompressing procedures, over the more simple cutting and partitioning ones. It merely stems from the essence of rescaling, in which the details may go away but the core elements of scaling objects persist and do not change as much as we see them in division or contracting, when a partial but irreparable loss of information is expected by default. So, in mathematical terms, one may expect the prevalence of exponentiation compared to the operations of multiplaying/dividing or adding/contracting.

Then, resuming the above, we enter the realm of binary logic with the set of possible outcomes of two in each consecutive step. Hence, the priority appearance of mathematical forms of type

$$T = 2^d \tag{7.1}$$

becomes an attribute of the simple physical world, where *d* is a number dealing with the topology of studying space, which in our case is a space of *SEE*, or just energy space (1.1.12.a).

3.1.4. *PoB* Landscape of Energy Space

As it was shown in section 2.2.1, the discrete nodes (2.3) have additional meaning as the points of bifurcation (*PoB*) that separate the qualitatively different areas in *SEE* space. The crucial thing here is that the role of the bifurcation points (2.3) is not the same.

To be exact, elaborated analysis confirmed the evident bifurcation essence of the first three spectral nodes (2.3), which are *GRP* (fundamental harmonic), *LSP* (second harmonic), and *GRE* (third harmonic), i.e., for $n \leq 3$. In physical connotation, *GRP* is the point of dynamic balance between energies invested in inheritance and variability processes, *LSP* is the point of probability parity between scenarios of advancement and degradation in *SEE*, and *GRE* is the point of dynamic balance between probabilities for the constructive and destructive interaction with energy flow in *OTS* [21].

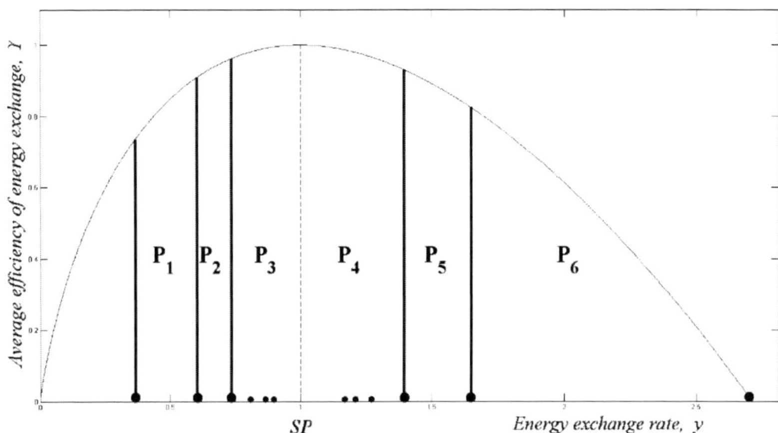

Figure 7.2. Discrete spectrum of energy exchange efficiency Υ in development of *OTS*. In the plot, the dependence of Υ (ordinate axis) on the rate of energy exchange y (abscissa axis) is shown. Nodes Υ_n are displayed by vertical segments (the rightmost Υ_n is zero and absent). In total, there are *6* phases referred to as *Pi*, where $i = 1, 2,... 6$. The significant *PoB* are marked by the bigger black dots, while the points with degenerate bifurcation behaviour are marked by the smaller ones concentrated around *SP*.

However, the nodes with $n > 3$ do not reveal bifurcation performance to as significant an extent as the nodes with $n \leq 3$ do. It may mean that although high harmonics keep their ordinary spectral properties, they do not bring tangible evolutionary changes, at least, utilizing standard apparatus of calculus.

Finally, notice that for exchange energy harmonics E_n in the k-basis at $y < SP$ holds

$$E_1 = \frac{1}{2} E_2 E_3 \tag{7.2}$$

which confirms the special meaning of this *PoB* area $\{E_1, E_2, E_3\}$ as a single whole.

So, the bifurcation behavior exists across the whole y-range $[GRP_L, GRP_R]$, although its manifestation changes with n. In turn, it assumes that *SEE* space is essentially under the control of the set of *PoB* that defines an infinite family of the internodal phases *Pi*, which will be considered in detail in the next section.

3.1.5. Segregation of *SEE* Space

Below, we describe consecutive steps in the unenforced rise of the separate phases in the *SEE* space as a consequence of the existence of the bifurcation points.

3.1.5.1. Primary Segregation, $d = (0 + 1) = 1$

Within the presented approach, the process of energy arrangement is generally accompanied by an unlimited number of random leaps of evolving object in the *SEE* space. On this journey, this object can land anywhere. However, if it lands outside the original energy range $[0, GRP_R]$, then, according to existing theory, its further presence in the *SEE* space as a unique unit is impossible. So, the existence of energy boundaries $y = 0$ and $y = GRP_R$ puts the primary cut and segregates the united energy range $[0, GRP_R]$ from the rest of the *SEE* space (Figure 7.2).

3.1.5.2. Secondary Segregation, *d = (1 + 1) = 2*

Remarkable *PoB* properties of fundamental harmonic GRP_L [64] create favourable conditions for dividing the original *y*-range $[0, GRP_R]$ into two parts: $[0, GRP_L]$ and $[GRP_L, GRP_R]$.

3.1.5.3. Tertiary Segregation, *d = (2 + 1) = 3*

Above, we have already compared physical conditions in *3* consecutive phases *Pi1-Pi3* based on Table 2.1 and came to the conclusion that the combination of properties in each *Pi* is unique. It motivates us to consider *Pi* phases as the areas of unlikely physical meaning. Hence, tertiary segregation splits both united stages $[GRP_L, GRP_R]$ and $[GRE_R, GRP_R]$ into *3* smaller dissimilar phases: *Pi1 – Pi3* and *Pi4 – Pi6*.

3.1.5.4. Quaternary Segregation, *d = (3 + 3) = 6*

We have already asserted the pair presence of *y*-points, so we may observe the energy clone of phase structure *Pi1-Pi3* in the area $[GRE_R, GRP_R]$. The existence of bifurcation behaviour (unlikely the one observed in *Pi1-Pi3*) in appropriate *y*-points was confirmed earlier in 2.2.1. So, we observe one more family of *PoB* as *Pi4 – Pi6* and declare existence in a total of six physically dissimilar phases, *Pi1 – Pi6* in the *SEE* space.

3.1.6. Math Behind Segregation of Space *y - Y*

From 3.1.5, it follows that we can consider the following steps in the transformation of an energy topology of originally uniform *SEE* space. These steps mean:

1. $d = 0$ - original uniform energy space $[-\infty, \infty]$ with no phases, no *GC* realizations.
2. $d = 0 + 1 = 1$ - emergence of one phase $[0, GRP_R]$ with *GC* realization 64^1 (App. C.1).

3. $d = 0 + 1 + 1 = 2$ - emergence of *2* phases *[0, GRP_L]* and *[GRP_L, GRP_R]* with *GC* realization 8^2 (App. C.2).
4. $d = 0 + 1 + 2 = 3$ - emergence of *3* phases *[GRP_L, LSP_L]*, *[LSP_L, GRE_L]*, *[GRE_L, SP]* to the left of *SP*, or, separately, *3* phases *[SP, GRE_R]*, *[GRE_R, LSP_R]*, *[LSP_R, GRP_R]* to the right of *SP*, with *GC* realization 4^3 (App. C.4).
5. $d = 0 + 1 + 2 + 3 = 6$ - emergence of *6* phases; *3* phases in the *y*-range *[GRP_L, SP]* and *3* phases in the *y*-range *[SP, GRP_R]* with *GC* realization 2^6 (App. C.3) as shown in Table 7.1.

Table 7.1. Consecutive split an original *SEE* space with the forming of adjacent phases

Segregation steps in *SEE* space				
Step's number, *n*	Number of phases, *k*	Number of nucleotides, *m*	*GC* realization	Level of space deformation, *d*
1	0	infinite	∞	*0*
2	1	8	64^1	*1*
3	2	4	8^2	*2*
4	3	2	4^3	*3*
5	6	1	2^6	*6*

So, for ordinal numbers *n*

$$1, 2, 3, 4, 5, \ldots \tag{7.3}$$

the consequence *k*

$$0, 1, 2, 3, 6, \ldots \tag{7.4}$$

comes as the Tribonacci pattern [22], which is a variant of the well-known Fibonacci sequence, starting with three predetermined terms, and each term afterwards is the sum of the preceding three terms, like below

$$T'(0) = 0,\ T'(1) = 1,\ T'(2) = 2,$$
$$T'(n) = T'(n-1) + T'(n-2) + T'(n-3) \tag{7.5}$$

Physically, it assumes that segregation of the *SEE* space takes a way of accumulation of change on each consecutive step. In other words, starting from an original undisturbed space with $k = 0$, each new segregation level superimposes on what was done before.

We desist from going further in consequences (7.3), (7.4) for the reason that the existence of the significant *PoB* with $n \geq 3$ is not confirmed.

The role of *6* as the maximum possible exponent *k* in the form 2^k is further discussed in section 3.1.8.

3.1.7. Physical Roots of *GC* According to *CEL*

As we see in 3.1.5, energy segregation of *SEE* space creates the basis for the existence of suitable physical constraints. It means that the status of an object (nucleotide) interfacing with segregated space will be evaluated by its relative position with respect to the existing energy infrastructure, i.e., phases *Pi*.

If an object does not hit the *Pi* boundary, its evolutional status does not change, no matter how many random leaps it has experienced in its evolutional journey to reach the current *y*-point. Obviously, crossing the *Pi* boundary causes an appropriate update in an object's status.

So, the evolutionary biography of a nucleotide and its current status is being written relative to its location within the existing energy infrastructure.

If the number of interfaced nucleotides *m* matches the number of available energy phases *k*, then different filling scenarios can still appear if the order of nucleotides does matter. Otherwise, if $m \neq k$, then the distribution of nucleotides over phases can take the form of a random combinatorial draw of *m* over *k*. In any relationship between finite *m* and *k*, there appears to be a limited number of unique filled positions in the *SEE* space. So, we come to a classic combinatorial problem for the placement of *m* nucleotides ("balls") over *k* phases ("bins").

It is worth noting that each energy phase during a new drawing may contain only one object, otherwise, it eliminates any meaningful drawing and creates a mess. Also, allow repetition of nucleotides in adjacent energy phases. Then, the permanent flow of *m* distinct objects through *k* distinct, unlikely phases will be filtered out, yielding a *k*-column matrix. So, in the essence of filtering objects by energy grid lies the physical properties of *PoB* as the points of critical change in the energy flows. So, the well-known trinucleotide's structure of the codon chain is a result of filtering the flow of *4* free nucleotides by the *3*-based (for realization 4^3) structure of *SEE*.

Now, let us revisit what was said so far in this chapter. On one hand, we have the consequence of free objects (nucleotides) in the *SEE* space. On the other hand, the ununiform *SEE* space dictates definitive rules (segregated energy phases) for any energy object in this space. Therefore, if the original consequence of objects follows these rules, it is unavoidably distributed over existing energy phases until object sequence or energy segregation exists. Then, we may consider such a road map for codon arrangement as a numerical energy skeleton of *GC*.

An example of such an energy skeleton for $m = 4$, $k = 3$, is shown in Figure 7.3 in the form of a *3 x 64*-entries table, in which each chart element fits one out of *64* possible codons [23] shown in Figure C.4. Underline that actually it is equivalent to the declaration on the triple-based structure of *GC*.

As we will see below, the described mechanism of *m-k* filtering may play an important role in a model for forming the energy skeleton of *DNA* as well (3.2.6).

3.1.8. Possible Origin of Number *64*

Now we are in a good position to evaluate the meaning of number *64,* as it follows from *CEL*.

So, on one hand, as follows from Section 3.1.6, the maximum factor of *SEE* space deformation $d = 6$. On the other hand, the maximum number of physically significant phases k in *SEE* space is *6* either as per 3.1.5.4, i.e., $d = k$ (further d and k are used interchangeably). So, we confirm by using two independent ways the existence of a maximum norm *6* in the *SEE* space, which in binary approximation yields

$$T_{max} = 2^6 = 64 \tag{7.6}$$

Based on considering context, $T_{max} = 64$ comes as an upper limit for the unique random realization of *GC* per one energy phase.

Hence, all physically permitted scenarios for *GC* realizations at $k \leq 6$ should respect

$$m^k = T_{max} \tag{7.7}$$

Based on (7.7), the effective number of nucleotides to support *GC* realization *m* depends on the measure of deformation of the *SEE* space *k*. As within the considering approach, the only integer-valued variants of *GC* realization are possible; then the integer roots of (7.7) are four ordered *m-k* pairs: *{2-6}*, *{4-3}*, *{8-2}*, and *{64-1}* that form the appropriate *GC* realizations 2^6, 4^3, 8^2, and 64^1, and will be considered in further text. In summary, we come to (7.7) as a constraint for the effective functioning of the sequence of nucleotides in segregated *SEE* space.

So, in considering theory, originally, the number of nucleotides is unlimited. The limitation steps in as a consequence of the segregation of *SEE* space (section 3.1.6). Then, *m* is the number of objects (nucleotides) to distribute over existing energy phases *k* (measure of deformation of *SEE* space), which finalizes the meaning and role of the number *64* (Figure 7.3).

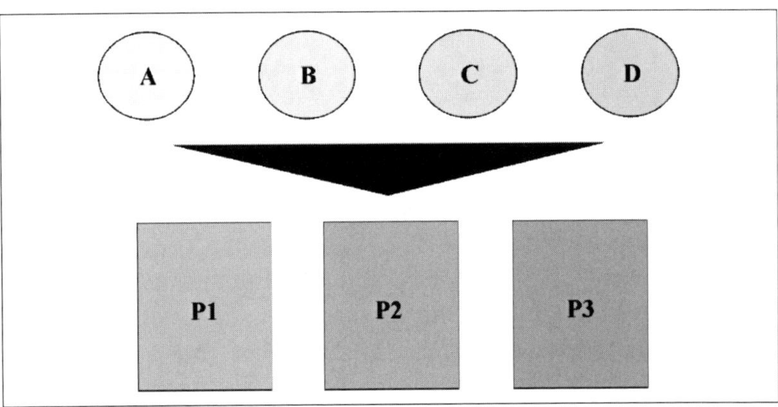

Figure 7.3. Schematic view for combinatorial placement of *4* "balls" over *3* "bins."

3.1.9. Combinatorial Basis for Calculation of *GC* Realizations

In the development of the discussed approach, the analysis conducted in App. *B* indicates variation as an optimal combinatorial method to deal with the *m* - *k* draw to hold an order (parameter of uniqueness of nucleotides' arrangement $\xi \neq 0$), whereas combination with repetition is an effective method to be applied for situations when the influence of order is negligible (parameter $\xi = 0$). Note that in this context the factor ξ has the obvious relation to the uniqueness of codon amino acids either; details are in section 3.1.12.

Then, at the realization of any hypothetical scenario based on (7.7), the number of possible outcomes respecting $\xi \neq 0$ is constant

$$V^k_m = m^k = 64 \tag{7.8}$$

where V^k_m is the variation for the total number of unique codon combinations.

However, the number of outcomes without order (parameter $\xi = 0$) varies and should be calculated for each realization separately as per

$$C_k(m) = \binom{m+k-1}{k} \tag{7.9}$$

where $C_k(m)$ is a combination of a k-th class of m elements [24].

Finally, as it is known [23], a degeneracy of GC refers to the fact that when amino acids are specified by more than one codon, then the degeneracy factor can be given as

$$D_g = V^k_m - C_k(m) \tag{7.10}$$

As by definition $V^k_m \geq C_k(m)$, then $D_g \geq 0$ and (7.10) works for any GC realization.

3.1.10. Possible Realizations of *GC*

Below, we give a short summary of each GC realization as it follows from (7.7). It is worth noting that if we fix the number of distinct nucleotides to the usual 4, then we should not expect any difference compared to the known results for the GC realization 4^3, no matter how many nucleotides in total participate in the draw. So, we are obligated to go out of the box and think that the number of distinct nucleotides can be not 4, even though it contradicts what we know for now.

The results of the simulation are presented in App. C. The most remarkable finding of App. C is that the realization 2^6 is possible only for the physical model of 1-strand DNA, i.e., for RNA (App. C.3). Note that, in reality, all GC realizations 64^1, 8^2, 4^3, and 2^6 probably coexisted and concurred at the same time, so realization 4^3 had to prove its efficiency. The common feature of all realizations is a degeneracy of the numerical structure. In conditions of real evolution, this fact is hardly the drawback, and sooner it is an advantage.

As a short summary, conducted analysis reveals two *GC* realizations that could be considered relatively mature without a significant chance for crucial errors. These *GC* realizations are 2^6 and 4^3. As it was mentioned, the major issue with 2^6 realization is the impossibility to come in a *2*-strand helix. Also, the double codon structure in the 2^6 realization leaves it as a rather archaic form of *GC* due to the plenty of possible outcomes required to support combinatorial dicing; it is as much as *384* as it is shown in the next section. Consequently, within the considering approach, the format 4^3 looks like an optimal solution. Yet, if the above is true, then realization 2^6 looks like the closest predecessor of modern *GC* realization 4^3, as is also discussed in 3.1.11.

3.1.11. Did Nature Make a Mistake?

Take a close look at the transitional chain between possible *GC* realizations

$$2^6 \leftrightarrow 4^3 \leftrightarrow 8^2 \leftrightarrow 64^1 \tag{7.11}$$

Pay attention to the obvious ratio between the consecutive items, i.e., $2^2 = 4$ and $8^2 = 64$; however, $4^2 \neq 8$. In other words, the transit from *4* to *8* cannot be accomplished using the same exponentiation pointer *2* but requires some odd (within considering approach) pointer *1.5*. It means that the rescaling at transit $4 \rightarrow 8$ cannot be realized within the claimed set of integers; therefore, transit lacks its consistency.

On the other hand, observe the strong asymmetry in the number of chart elements (supporting outcomes) for declared realizations: the realization 8^2 uses $64 \times 2 = 128$ elements (Figure C.2), leaving $64 \times 3 = 192$ elements for the realization 4^3 (Figure C.4), and high $64 \times 6 = 384$ elements for the realization 2^6 (Figure C.3). Obviously, the above differences are the factors in the evolutionary process, which directly calls for efficiency of the used realization [62, 63]. Evidently, the higher volume of outcomes at the decrease of used nucleotides puts the extra evolutionary pressure on the realization mechanism and leads to the potential increase in the error rate in the protein's assembly.

That is why once the jump from 8^2 (*128* outcomes) to 2^6 (*384* outcomes) was done and nature faced the consequences of the higher error rate, it may have taken a step back and implemented the 4^3 realization with *192* outcomes by just as simple a squaring of *2*.

Though, perhaps, nature set the realization 4^3 as a goal at once and just did an outflanking manoeuvre to obtain 2^6 at first, from which to make an ordinary exponentiation of *2*. So, as in mathematical terms, the 4^3 realization cannot be consistently obtained from the 8^2, we may think of the transition

$$8^2 \rightarrow 2^6 \rightarrow 4^3 \tag{7.12}$$

as "nature trickiness" (Schema 7.1).

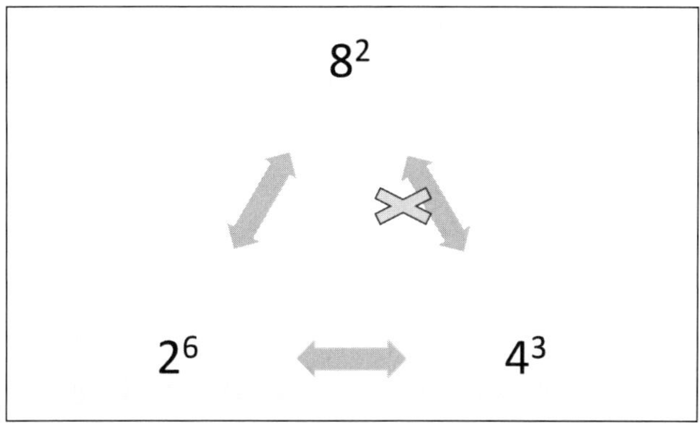

Schema 7.1. Possible consequence of *GC* realizations in evolution.

Note that if above is not far from true, then realization 2^6 comes as the closest predecessor of modern *GC* 4^3.

3.1.12. *GC* Structure from *CEL* Standpoint

As it was mentioned above, in this book, the author did not have the plans to analyse the known structure of a *GC* and investigate combinations of nucleotides or codons capable of making a difference in *GC* functioning. Instead, the presented narrative is aimed at studies of the possible origin of the current revision of *GC*, in part due to the fact that the major author's tool, the model *CEL*, does not provide any trustable mathematical instruments to handle biochemical peculiarities of existing *GC*.

Keeping that in mind, nevertheless, the author reserves the right to express his comments on a few things related to *GC* structure that may have rational roots in *CEL*.

To track the origin of *GC*, we followed the evolutionary path forwards from the starting point to what we know about the structure of *GC* for now. So, by taking a retrospective look at this path, it comes that an energy skeleton of *GC* is the natural algorithm of the interaction between free nucleotides and the energy infrastructure of *OTS*.

Now, we intend to compare the known observables of *GC* with what the *CEL* can provide. To do that, below we divided the statements normally prescribed to *GC* into two groups. The first (*I*) group includes the properties that can be directly supported by *CEL*, while the second (*II*) group lists the properties that cannot be understood within *CEL*. We restrict our analysis to the master realization 4^3 only. Also, it is worth noting that we did not discover the properties that directly contradict *CEL*.

Group *I*:

1. Triplet structure: *CEL* provides an intuitive way to support the triplet-based structure of *GC* through the triplet structure of the *SEE* space (phases *Pi1-Pi3* and *Pi4-Pi6*) declared in section 3.1.5.3.
2. Transition from a *64* codons alphabet to a *20* amino acids alphabet: number *64* comes from (7.8) as the total number of codons respecting the requirement of codon's order $\xi \neq 0$. However, to codify the unique amino acids, we should count the unique codons at $\xi = 0$. Then, from (C.4), the number of necessary amino acids is *20*.
3. Nonoverlapping *GC* code (a single nucleotide cannot be part of two adjacent codons): In *CEL,* it means that an energy phase includes more than one nucleotide in each evolutionary step, which is not possible.
4. Comma-less structure of *GC* (no room for punctuation in between codons): Also follows from the nonoverlapping essence of *GC* above.
5. Degeneracy (a few codons may correspond to the same amino acid): Aligned with (7.10).
6. Non-ambiguity (each codon corresponds to a particular amino acid only): Confirmed, as otherwise the requirement for *20* amino acids is violated.

7. Absence of codons for *D*-amino acids [25] able to bring two particular stereochemical formulae into coincidence in space: In *CEL*, original absence or presence of *D*-amino acids is not required. The *L* amino acids can act as a substrate, perhaps an energy substrate, in *D* amino acid synthesis. The *L* and *D* amino acids have different energy footprints. Conversion from *L* to *D* can happen subject to changes in the energy environment at $GRE_L < y < GRE_R$.
8. Presence of signal codons (start- and stop-codons): In *CEL*, the existence of the signal codons is a natural process of regulation of the rate of chemical reactions in an energy exchange. The details are in sections 3.1.13 and 3.2.6.

Group *II*:

1. Collinearity (codon sequence fits to amino acids sequence in coding protein).
2. Universality (the same genetic code is valid for all the organisms).
3. Unidirectionality (proteins reading takes a sequential way with no misses or returns).
4. Fixed polarity of codons (each triplet is read from the 5' → 3' direction).
5. Limitation of synonymous replacements by third position of codons.

3.1.13. Composition of Signal Codons

It is reasonable to think that signal codons, such as stop and start codons, are the sort of control sets that essentially regulate the process of protein assembly to minimize the risk of ambiguity in *GC* translation. Take a close look at what we know about the structure of signal codons through the prism of *CEL* theory using the traditional transcription of *RNA*'s nucleotides *A, C, G,* and *U* [26].

In this section, we again focus on the 4^3 realization only, as up until these days, genetic society has accumulated a high volume of appropriate knowledge on this subject.

Absence of *C* codon.

So, the complete absence of nucleotide *C* within the structure of signal codons in *CEL* language may signify the relative insignificance of *C* compared to the contribution of the other *3* nucleotides to total *SEE* performance. As it follows from Figure 7.1, the absorbance profile of *C* is the smoothest one, and its peak is the very smeared. In other words, it corresponds to the minimal rate of absorbance change, which makes it the most vulnerable part in terms of proper identification and blurring.

Stop codons and *PoB*.

In *CEL*, the number of stop codons is exactly equal to the number of *PoB* points before and beyond *SP*. It could mean an association between the termination of some stage in *SEE* and generating a signal to terminate the process of protein assembly, especially accounting for the critical role of *PoB* as the point of reconfiguration of energy flows (section 3.2.6).

Comparison of structure in signal codons.

Below, we use the standard table of *64* different codons in the genetic code [27].

Compare the structure of the start codon *AUG* with the group of stop codons *UAA, UAG,* and *UGA*. We see that the two-element fragment *AU* in the start codon changes to an opposite *UA* in the nucleotide order in the strongest stop codons *UAA, UAG*, or *UG* in the weaker *UGA*. It may mean that the *UA* rather takes the role of an effective damper of assembly mechanisms in contrast to *AU*, which, appropriately, is an effective igniter.

Based on that, consider possible compositional forms of the signal codons. Firstly, as we excluded *C*, then we have to exclude from the general list the codons *UAC* and *AUC* either.

Next, pay attention to codons *UAU* and *AUA* that are followed by consecutive substitutes of *U* and *A*. The internal structure of these codons may come internally contradictory if presented as *A(U) + ... + (U)A* and *U(A) + ... + (A)U*, correspondingly, where transitional nucleotide is taken in brackets. Then, the two-letter commands to stop (effective damper *UA*) and to start (effective igniter *AU*) may cancel each other. As a result, we may remove codons *UAU* and *AUA* from the list of signal codons.

Summarizing, the real stop codons *UAA* and *UAG* are the only possible formats for the stop codons, while *AUG* and *AUU* (hypothetically) are the acceptable forms of the start codons.

3.1.14. Final Remarks

According to an official definition, *GC* is the rule for translating the *64*-letter codon alphabet to the *20*-letter alphabet of amino acids [28]. The presented chapter is an attempt to work out an answer to the possible mechanism for the origin of this translation based on the central for *CEL* idea of *SEE*. To do that, it was suggested starting from scratch on the front line of interaction between energy flux and an open system. Then, going this way, ultimately, it was shown that *GC* can be understood as the product of the interaction between the sequence of nucleotides and the infrastructure of the energy exchange between *OTS* and the environment.

Notice that the existence of the consequence of free nucleotides as an essential element of the formation of *GC* was highlighted by many authors. In this Chapter, only one crucial element was added to this puzzle: a natural three-part energy filter positioned between the sequence of nucleotides (adapted nucleotides) and the energy place for assembly of *DNA*.

Notice that discussion about the phenomenon of discreteness (energy phase separation in our case) in genome organization across evolution is a regular process; in confirmation, mention just a few works [29-31]. The authors [29] stress that evidence for phase separation exists across the entire tree of life, while [30] underlines the principally stochastic nature of genome organization and function. Specifics of charge separation in the disordered region are discussed in [31].

In turn, the infrastructure of the energy exchange follows the discrete pattern of the *CEL* energy solution due to the confinement of energy exchange within the natural limits imposed on the rate of energy and entropy variation. Additional research revealed that the three first nodes of the discrete solution are unique, in contrast to the remaining nodes, are at the same time, the *PoB* and the points of dynamic balance (3.1.4). Later, the falling of energy positions of *GC* signal codons into the same energy zone was confirmed (3.2.6), which set up the "energy footprint" between the *SEE* focused chapters in Part 2 and the current one devoted to the origin of *GC*.

Generally speaking, in the considered theory, the process of *SEE* is under permanent pressure from the factor of energy segregation. Evidently, any limitation of chaotic performance imminently affects the character of physical and, therefore, biological processes [32-34]. The root reason for such a phenomenon lies in the necessity of adapting to emerging energy rules. Through existing physical laws, it unavoidably imposes the more or less serious restrictions on the regular function of all entities. In this way, one of the late outputs of the segregation pressure is the plan for the evolution of *OTS*, formally put in the form of a *64*-codon matrix.

Thus, due to existing laws of nature, to be continued, an energy development has to pre-encode itself in some numerical energy vocabulary. In light of this, energy evolution comes as a physically accommodating process that transfers energy from the local chaotic environment through segregation and pre-assignment of phases of chaos to stable chemicals [35], and this is what we may call an "energy code" [36, 37] implemented through the scenarios put in practice by material carriers. Due to existing energy limitations, such encoding takes the form of a random realization according to combinatorial rules.

Another crucial aspect of energy development quantified in this research is the degree of concurrency between the factors responsible for the conservation and change of *OTS* energy structure. It was shown that, in essence, variation and inheritance are natural attributes of an *SEE* process. On the one hand, they demarcate the *SEE* process, while on the other hand, they support the process of fluctuations, thereby making sure the permanent flow of energy opportunities for *OTS* needs exists. In this sense, it becomes clear that variation-inheritance coupling may serve as a forerunner for the driving force in natural evolution.

It is worth noting that, by default, the investigated structure of *GC* belongs to the actual winners in the evolutionary competition that has started on primordial Earth. Admittedly, all probabilistically significant deviations from existing *GC* were ultimately suppressed millions of years ago, and we cannot say for certain what the structure of *GC* in these extinct carriers has been. That is why, to avoid speculations, we have deliberately kept only a short discussion on possible relatively predecessors of modern revisions of *GC* realization and focused on the interpretation of what we know for sure now.

In some approximation, if we ignore the obvious difference in scale, the above process of continuous to discrete transformation resembles the well-known development of tree-ring segregation. Firstly, the formation of an individual pattern reflects changes in the local climate, predominantly temperature, and secondly, the possible way to retrieve these climate changes is to properly decipher the existing tree-ring pattern. At this point, it is worth noting that the dividing of tree cells follows a highly controlled sequence of successive events described in the cell cycle. So, the emergence of the discrete solution in this example looks inescapable [38]. Hence, in the reverse way, if *GC* is fully understood and translated, it should reproduce a pattern of the sophisticated way for energy evolution in all its diversity, no matter whether it is the tree-ring segregation or the process of evolutionary changes.

The novelty of this report, predominantly, lies in the investigation of a general physical and mathematical background that can support the natural transformation of a chaotic energy flow through primitive *OTS* into a structured biochemical response. In this approach, the great impact on the energy development of *OTS* has the recently confirmed phenomenon of infrastructure in evolution, especially the triplet-based structure of *OTS*'s interface.

In other words, we have traced how sufficiently general and universal reasoning about the configuration of energy exchange can cause the formation of a structure of the codon chain and meet the challenge of numbers *3*, *4*, *20*, and *64*. Intercoupling between the magazine of available nucleotides and bifurcation phases creates a playground in which the combinatorial methods formalize the rules of the game.

In conclusion, the above results give a key to understanding fundamental mechanisms that (1) form numerical structure in pre-biological arrangements and (2) govern the development of these arrangements through concurrency between factors of variability and inheritance. The *64*-scenario matrix serves as the basis for the practical realization of a random algorithm in energy evolution and thereby plays the role of the workhorse for the *GC* algorithm. Nature drives this matrix repeatedly, producing diverse energy patterns and creating codon chains from the very controlled number of available elements. So, *GC* surprisingly works in the continuous struggle with intrinsic noise while accurately and efficiently translating discrete inheritance information no matter what [39]. The presented theory provides a unified framework for the origin and mathematical structure of *GC*.

Conclusion of Chapter 3.1

1. Given the existence of the flow of nucleotides, the arising of *GC* appears a natural product in evolution of *OTS*.
2. To defeat chaos and maintain the uniqueness of individual properties, *OTS* supports the emergence of the *GC* as a limited, discrete set of highly protected units.
3. The *OTS*, probably, picked up a rescaling (exponentiation) over other mathematical operations (multiplying, division, addition, contracting) as the most proper mechanism for the transfer of sensitive hereditary information.
4. The binary logic in each consecutive step of evolution comes as the preferable mathematical toolset in evolution of *OTS*.
5. *PoB*-deformation and segregation of original uniform energy space is an intuitive result of *SEE*. Significant bifurcation performance is within the range $[GRP_L, GRE_L]$ (at $n \leq 3$).

6. Energy segregation in *SEE* space goes through an accumulation of change on each consecutive step, when each new segregation level superimposes on what was done before. The result of such segregation is the emergence of energy phases.
7. In the chaotic evolution, an interaction of the sets of distinct nucleotides and energy phases, both of non-equal volume, looks only possible on the combinatorial basis.
8. Within the existing *4*-nucleotide – *3*-phase concept, the physical meaning of the number *64* is the upper limit for the unique random realization of *GC* per one phase, which yields *4* possible *GC* realizations, which in ascending efficiency order are: 2^6, 4^3, 8^2, 64^1.
9. Two *GC* realizations without a significant chance for crucial errors are 2^6 and 4^3. The major issue with 2^6 realization is the impossibility to come in a *2*-strand helix. Then, the format 4^3 looks like not the best but rather an optimal solution.
10. *GC* does not demonstrate the properties that directly contradict *CEL*.

Chapter 3.2

The Energy Skeleton of *DNA*

It is hardly possible to fight with an idea that the origin and evolution of *DNA* and *RNA* (further *DNA*) occurred with the participation of physical laws, or even more, evolutionary transformation was directly governed by the physical regularities. In this sense, the appearance of *DNA* is a product of the complicated and deep influence of the known and maybe not well-understood yet physical concepts accounting for that typically energy relations work in the background.

That is why, based on the content of the above chapters, below we investigate the possibility for an origin of the energy skeleton of *DNA* as a natural result of the employed paradigm *CEL*, in which the advent of *DNA* is expected to be a direct consequence of the structure of *SEE* discovered in Part 2.

So, the purpose of this chapter is to present theoretical construction, which, on one hand, logically follows results obtained earlier and, on the other hand, demonstrates basic properties of real *DNA*.

No doubt that *DNA*, as any other object of the real physical world, can be considered under different angles. Further, we will touch just a few of them that constitute necessity and inevitability for the appearance of such a physical construct as *DNA* reflecting the logic of cell energy development. Hopefully, the suggested view on the occurrence of *DNA* will help to piece together the mosaic of many energy exchange processes in the cell into one non-contradictory framework, keeping the advantages of the discussed high-level view at the *SEE* transformation.

In line with the above-mentioned, all further research will be done based on the previous results obtained while developing *CEL* methodology.

3.2.1. Forming of Complementary Antiparallel Strands

Consider whether *CEL* is capable of providing any mechanism in support of the energy integrity of the found *SEE* solution (6.7).

In this connection, recall that the permanent flow of nucleotides in a permanent interaction with the energy phase structure of *OTS* unavoidably distributes nucleotides over energy phases (section 3.1.7) on the basis of one nucleotide per one phase (3.1.12), otherwise, it breaks a principle of nonoverlapping *GC*. So, in the maximum case possible, the *6* phases are available for nucleotide residence, which causes the appearance of *2* sets of phases, the left one $[GRP_L, SP]$ and the right one $[SP, GRP_R]$. Based on that, two originally independent *y*-adjacent sets (strands) with the speckles of nucleotides (call one template, another one complementary) can appear [40-42].

Highlight that the sense of circumnavigation of contours in these consecutive strands is opposite, i.e., the right-hand (clockwise) sense in the range $[GRP_L, SP]$ and the left-hand (counterclockwise) sense in the range $[SP, GRP_R]$; in other words, the strands are located in the *y*-ranges of an opposite polarity (1.10.a,b). At that, the counterclockwise sense in the range $[GRP_L, SP]$ and the clockwise in the range $[SP, GRP_R]$ are prohibited (Figure 1.1).

Next, while in the template strand $[GRP_L, SP]$, the natural course of *y* is from the left to the right; in the complementary strand $[SP, GRP_R]$ the situation is opposite, and the natural change of *y* to reach *SP* is opposite from the right with $y = GRP_R$ to the left with $y = SP$. It provides grounds to conclude that the considering strands run in the opposite directions, and they are antiparallel (Figure 8.2).

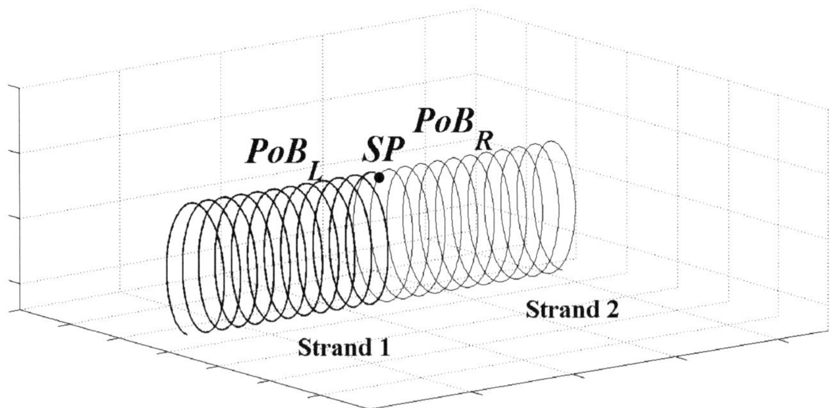

Figure 8.1. Schematic view of *2* consecutive energy strands (helixes). The left (template) one $[GRP_L, SP]$ is shown in bold font, and the right (complementary) one is shown in regular font $[SP, GRP_R]$. PoB_L and PoB_R denote the sets of *PoB* on the left (PoB_L) and on the right (PoB_R) of the point *SP*.

Also, from (2.17) follows that the y-points in the ranges *[GRP_L, SP]* and *[SP, GRP_R]* meet the requirement of complementarity.

Summarizing, the energy strands *[GRP_L, SP]* and *[SP, GRP_R]* are complementary to each other, demonstrate an antiparallel course with the opposite sense of circumnavigation, and reveal a linkage between the paired points (Figure 8.1). So, to the best existing understanding, we may claim that the *CEL* energy strands demonstrate a proximity in the properties to real *DNA* strands.

3.2.2. Phenomenon of Energy Affinity in *CEL*

Consider in slightly more detail another aspect of the *CEL*-solution related to the opportunity to achieve the internal energy stability for configuration 8.1.

Due to obvious reasons, *CEL* is not capable of handling the chemical affinity between nucleotides. At the same time, *CEL* has a physical mechanism to support an energy affinity between the y-phases, where nucleotides can reside. This mechanism is highlighted below.

So, in accordance with the phenomenon of logarithmic inverse (1.2.6), any y-point in the range *[GRP_L, SP]* has its own not linearly symmetric replicate in relative to the line $y = SP$ in the range *[SP, GRP_R]*, i.e., all y-points in the ranges *[GRP_L, SP]* and *[SP, GRP_R]* come up only in pairs (2.17).

Recall that y in its physical meaning is the genuine energy quantity, which brings (2.17) as an energy relationship. At regular changes of y, due to (2.17), appropriate changes in one paired y-point should cause suitable changes in the other paired y-point. In other words, the paired points are constantly exchanging energy with one another, thereby supporting the energy flow between them. Evidently, maintenance of energy stability requires balance between energy streams that, in that case, may be achieved through adjustment in oppositely directed point-through flows.

Then, the *CEL* strands cannot be thought of as independent any longer; instead, they are now bound by the mechanism of energy pairing, and the result of this mutual dependence will be unavoidably fostering to the changes in the energy structure of connected strands. Therefore, in *CEL*, the existence of the energy affiliated paired points provides solution (8.1) with more energy stability.

Highlight here that due to the logarithmic essence of presented symmetry, the strands *[GRP_L, SP]* and *[SP, GRP_R]* are not completely congruent. On the contrary, the logarithmic symmetry assumes the existence of the constant shear stress that, altogether with the antiparallel character of the strands may cause the non-zero rotational angular momentum, which enforces the strands to twist in relative to the axis $y = SP$ and each other. Accounting for the above geometrical asymmetries, similar twisting of two strands with the same axis hardly causes the appearance of *2-D* effects only. Far more likely is the emergence of the third dimension and the ultimate forming of *3-D* spatial configuration.

So, in *CEL*, the template and the complementary strands are connected to each other by energy links that, firstly, provide more energy stability and, secondly, create a twisting moment that may contribute to the forming of a *3-D* structure composed of two strands with a common axis (Figure 8.2).

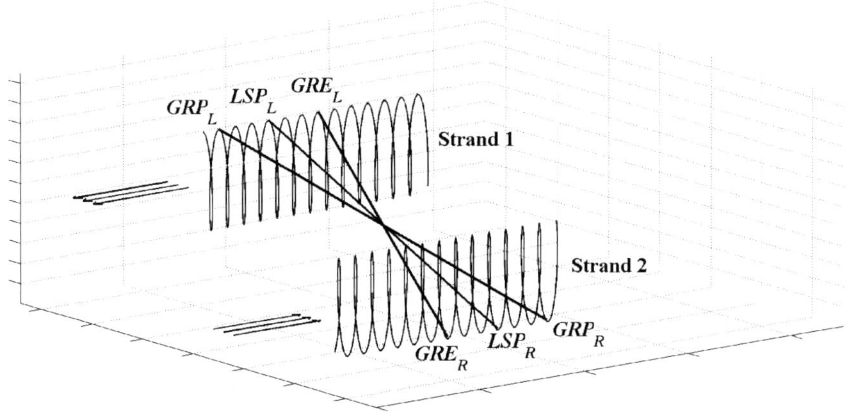

Figure 8.2. Schematic view for the forming of a *3-D* double energy helix at the emergence of pairing energy links. The *PoB* on Strand *1* (GRP_L, LSP_L, GRE_L) and Strand *2* (GRP_R, LSP_R, GRE_R) are shown. The energy connections between Strands *1* and *2* are shown by bold segments.

3.2.3. Channelling of Energy Exchange

Multiply both parts of the *SEE* equation (6.7.a) by π, then

$$\pi(E^C_{n\,\min})^2 + 2\pi\hat{E}^T E^C_{n\,\min} - \pi(k_n\hat{E}^T_n)^2 = 0 \tag{8.1}$$

where E_n is the *n-th* harmonic of the energy spectrum (2.3).

Pay attention that based on Chapter 2.2, the quantities $\pi(E^{C}{}_{n\,\min})^2$, $\pi(k_n\hat{E}^T{}_n)^2$, and $2\pi E^T{}_n E^C{}_{n\,\min}$ can be interpreted as suitable areas. Then, introduce $S_S = 2\pi\hat{E}^T_n E^C{}_{n\,\min}$ as the side area, $S^A{}_{n1} = \pi(E^C{}_{n\,\min})^2$ and $S^A{}_{n2} = \pi(k_n\hat{E}^T_n)^2$ as the area of appropriate circles with radius $\hat{E}^C{}_{n\,min}$ and $k_n\hat{E}^T{}_n$, correspondingly, so observe the quadratic form

$$\Delta E_n{}^2 = S^A{}_{n1} + S_{nS} - S^A{}_{n2} \tag{8.2}$$

In form (8.2), circinate the side surface S_S to accommodate $S^A{}_{n1}$ and $S^A{}_{n2}$, then we obtain a *3-D* quasi-cylindrical shape (Figure 8.3), where the circle of radius $k_nE^T{}_n$ ($E^C{}_{n\,min}$) is translated along the axis of figure (8.3) with the horizontal raise segment $E^C{}_{n\,min}$ ($k_nE^T{}_n$). Generally, $S^A{}_{n1} \neq S^A{}_{n2}$, and S_S does not fit the curvature of $S^A{}_{n1}$ and $S^A{}_{n2}$; therefore, we must figure out (see next section) whether $S^A{}_{n1}$, $S^A{}_{n2}$, and S_S are capable of creating the unbroken one-piece figure.

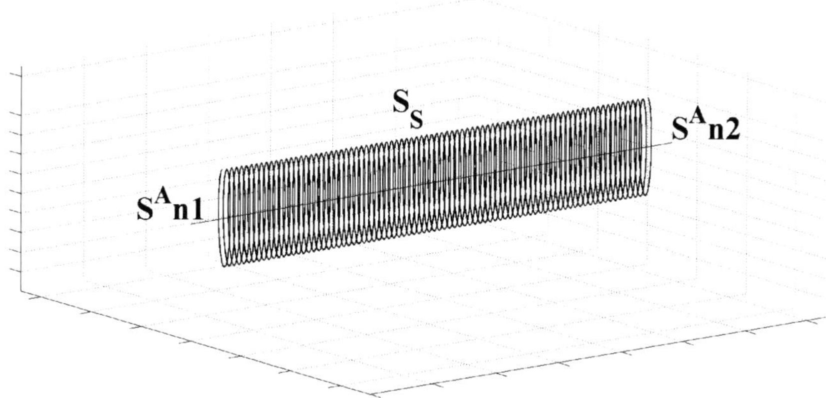

Figure 8.3. *3-D* quasi-cylindrical shape in *SEE* space dealing with the energy image of *SEE*. The S_S, $S^A{}_{n1}$, and $S^A{}_{n2}$ are the quantities dealing with an area in the *SEE*. So, $S^A{}_{n1}$ and $S^A{}_{n2}$ are the areas of the circles with radius $\hat{E}^C{}_{n\,min}$ and $k_n\hat{E}^T{}_n$, whereas $S_S = 2\pi E^T{}_n E^C{}_{n\,\min}$ is the side area.

From the physical viewpoint, (8.1) by its origin (6.7.a) is a relation of energy exchange, and it deals with the balance of energy flows (guided energy transports) in the *SEE* space, where, in line with the meaning of E^C and E^T discussed in the section 2.3, the areas $\pi(E^C{}_{n\,\min})^2, \pi(k_n\hat{E}^T{}_n)^2$, and $2\pi E^T{}_n E^C{}_{n\,\min}$ serve bidirectional transfer for energy supporting competition between the factors of inheritance ($E^C{}_{n\,min}$) and variability ($k_nE^T{}_n$).

3.2.4. Condition of Perfect *3-D* Energy Cylinder

To answer the question raised in the previous section about the compatibility of $S^A{}_{n1}$, $S^A{}_{n2}$, and S_S, we must resolve

$$\Delta E_n^2 = 0 \tag{8.3}$$

Then, as shown in Chapter 2.3, the solution of (8.3) is

$$\eta_{n_{1,2}} = -1 \pm \sqrt{1 + k_n^2} \tag{8.4}$$

where ratio

$$\eta_n = \frac{E^C{}_{n\,min}}{E^T{}_n} \tag{8.5}$$

Recall that from (2.14), factor k is limited by the range *(0, 2]*, which leaves the η-roots within two ranges, $\Delta_{\eta 1} = [0,\ 1.23]$ and $\Delta_{\eta 2} = [-3.23,\ -2.0]$, shown in light grey in Figure 8.4. The above limitation has an obvious physical reason, as it keeps the solution within the *y*-range *[GRP$_L$, GRP$_R$]*, where physically permitted solutions exist. Note that dependence $\eta = \eta(k)$ was analyzed in section 2.3.1 and presented in Figure 6.1.

Hence, the quantities $S^A{}_{n1}$, $S^A{}_{n2}$, and S_S can be made compliant if

$$\eta \in \Delta_{\eta 1}, \Delta_{\eta 2} \tag{8.6}$$

In this case, (8.1) describes a geometrically perfect cylinder (Figure 8.3).

If (8.6) violates, it moves the solution up to the highly chaotic area of the strong gradient ΔE^2 (intense energy fluxes) with $k^2 > 4$ or the predominance of energy decay ($k^2 < 0$). As a result, outside (8.6), any energy quasi-balance can be achieved only at the expense of intensive energy reconfiguration, which makes the stability of the achieved state vulnerable and, therefore, hardly possible.

Below, focus on the range $\Delta_{\eta 1} = [0, 1.23]$. Before continuing analysis, highlight that any energy process can be identified by its footprint in terms of its projection onto the plane η - ΔE^2 (Figure 8.4). As it is seen, some footprints (curves in the η - ΔE^2 space) cross the areas of both signs, i.e., $\Delta E^2 > 0$ and $\Delta E^2 < 0$, while others are located within the one sign area, keeping ΔE^2 only positive or only negative.

Bearing this in mind, in the plot, draw a few ΔE^2-curves by playing with factor k in (8.4). Put focus on the curves with k fitting to the significant *PoB* that are shown in Figure 8.4 with different line styles. Further, call these curves collectively the *PoB*-curves, and individually the GRP_L-curve, LSP_L-curve, and so on.

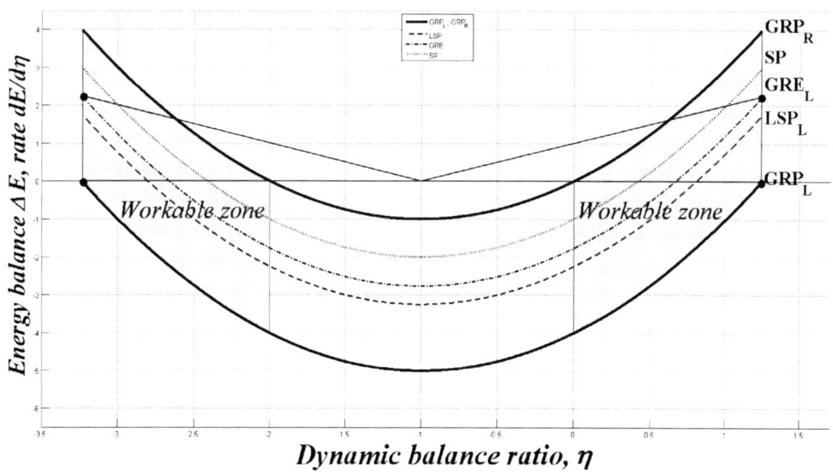

Figure 8.4. Plot of energy balance $\Delta E^2(\eta)$. The workable energy sign-changeable zones are shown in light grey. Within the workable zones, an effective sign-changing regime of ΔE^2 is in effect. The remaining plot area does not support a sign-changeable regime for ΔE^2. Two pairs of bold points in the left- and rightmost areas mark the intersection of the secants $\pm C(\eta+1)$ with curve $\Delta E^2(\eta)$, where $C = 0, 1$.

Pay attention to the positioning of the *PoB*-curves with regards to the segregation line $\Delta E^2 = 0$ within the coloured zones. While the LSP_L-, GRE_L-,

SP-, and other curves cross both the positive ($\Delta E^2 > 0$) and negative ($\Delta E^2 < 0$) areas, the GRP_L-curve (and curves with greater k) and GRP_R-curve (and curves with less k) are exclusively located outside the coloured area.

Therefore, as outside the coloured area the sign ΔE^2 does not change, it prevents the full-fledged energy exchange in this area and reduces it to the plain sign-keeping regime of only receiving or only giving up energy. On the contrary, the favorable conditions to support energy exchange exist only within the range of alternating sign of ΔE^2, i.e., in the coloured areas shown in Figure 8.4. Notice that geometrically the coloured areas are made up of identical curvilinear triangles providing equal probability for PoB-curves to occupy the parts of opposite sign of ΔE^2.

In this sense, forming the accurate geometrical response is only possible within the coloured regions and the right energy cylinder (8.3) can exist only within the sign-changeable area of ΔE^2.

3.2.5. *PoB* Zone of Effective Energy Exchange

Compose the family of secants

$$\frac{\partial(\Delta E_n{}^2)}{\partial\eta} = -C(\eta+1) \tag{8.7}$$

In this family, only two lines with an integer C intersect the colored area: one with $C = 0$ (corresponds to GRP_L) and another one with $C = 1$ (corresponds to GRE_L).

Pay special attention to two $\{\Delta E_n{}^2,\eta\}$ points of the triple intersection marked in Figure 8.4 by bold. The one is paired point with coordinates *(1.23, 0)* or *(1.23, 4)*; the other point is *(1.23, 2.23)*. Firstly, these points are intersections of C-secants with the vertical cutoff line $\eta = 1.23$. Secondly, they are the intersections between the cutoff $\eta = 1.23$ and the PoB-curves, the GRP_L-curve and GRE_L-curve. Thirdly, they come as a result of the intersection between the C-secants and the abovementioned PoB-curves.

Obviously, the reported triple intersections are hardly just a freak of chance; moreover, if we are talking about such an indispensable element of the SEE structure as PoB. What may convince even more is that in addition to the above findings, these PoB are the points of a dynamic balance as they directly depend on number φ (D.1.a, D.1.b), which is a sufficiently significant fact in itself. These points are the η - ΔE^2 images of dynamic balance in the y – Υ phase space corresponding to $k = 2$ and $k = 4/3$, appropriately.

The important thing is that typically a dynamic balance point is a clear indicator of qualitative change (border effect) in the concomitant processes. In this context, the points $y = GRP_L$ and $y = GRE_L$ rail off the areas of different physical meaning, which are the area of constructive interaction of *OTS* with an energy flux ($GRP_L \leq y \leq GRE_L$) and the remaining area of more chaotic physical conditions.

The segregation essence of the points GRP_L and GRE_L was discussed in Chapters 2.1 and 2.2. For GRP_L, it is demonstrated in Figure 5.1. The point GRE_L in the considering context comes as the well-known Reynolds number R_E with regards to energy flow (Table 5.1), where $R_E = dT/dy{:}dW/dy$. The quantized appearance of R_E is shown in Figure 8.5, where the maximum possible value $R_E = 4$ (5.6) corresponds to $y = GRE_L$. So, the border character of GRE_L thereby is also exhibited.

Notice that above also attests to our earlier inference on the significance of the *PoB* with $n \leq 3$ only (3.1.4), which in the newly introduced terms is equivalent to a declaration on the occurrence of an efficient energy interaction with energy flow within the comparatively narrow y-range $(GRP_L, GRE_L]$.

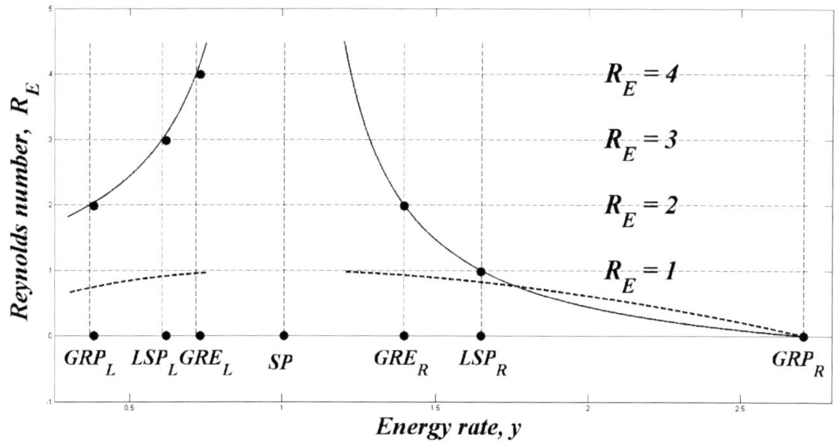

Figure 8.5. Energy profile of Reynolds number for energy flux. It is seen that the *PoB* points match the integer Reynolds numbers of *0, 1, 2* (twice), *3*, and *4*.

So, conducted reasoning on the coincidence of the locations for the area of significant *PoB* (dynamic balance points) and the sign-changing area of ΔE^2 makes deep physical sense. It indicates that the existence of a physically meaningful solution (8.2) and the formation of stable energy constructions in the *SEE* space is mostly probable within the range of significant *PoB* at $n \leq 3$.

3.2.6. Location of Energy Barriers and *DNA* Growing

In the previous sections, it was confirmed the role of the significant *PoB* in *CEL* theory as the critical points, demonstrating marginal behaviour and, as a rule, separating the ranges of opposite physical value. In this interpretation, the physical meaning of *PoB* is close to the normally understood role of energy barriers that are the routine feature of the energy environment in cells and work as effective regulators for the rate of chemical reactions [43].

Then, we may think that similarly to the preferred location of *PoB*, the energy barriers dealing with adjustment of energy exchange are also more probable within the energy range $y = (GRP_L, GRE_L)$.

On the other hand, it was settled that the *CEL* solutions (8.1) as energy configurations (*DNA* conformations) are probabilistically attracted towards the area of the sign-alternating energy exchange (section 3.2.5). Then, it looks natural to suppose that *DNA* conformations are also more probable within the area of the sign-alternating energy exchange matching the borders of the *PoB* area, i.e., it is assumed the interconnection between the energy position of *PoB* and *DNA* conformation that assigns a location of *DNA* conformations in the same *y*-range $[GRP_L, GRE_L]$ as shown in Figure 8.6.

Before we continue this line of thinking, it is worth a brief digression on the likely energy meaning of the *GC* signal codons. The critical role of the stop-codons as terminators of the protein's synthesis is well known [44, 45]. At the same time, the stop-codon itself, by its literal meaning as a carrier trinucleotide sequence of *DNA* capable of the transfer of chemical energy and an element of the cell signaling system, does not have any meaningful energy-related position. On the other hand, the formation of *GC* in *CEL* theory is a result of the interaction between the sequence of nucleotides and the triple-phased energy structure of *OTS* within the *y*-range $[GRP_L, GRE_L]$ (section 3.1.7). Therefore, though the nucleotides by themselves do not have recognizable energy signatures, the result of their interaction with the *OTS*'s energy structure does. As a result, the *GC*'s adapted nucleotides possess a clear energy identification mark as per their hosting phases do. Hence, it can be stated that the stop-codons are projected to the same *y*-area $[GRP_L, GRE_L]$.

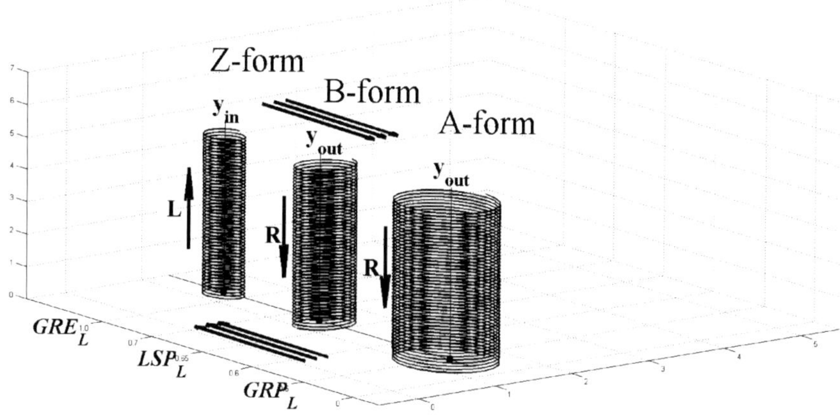

Figure 8.6. Schematic view of the major conformations of *DNA*. It is shown that the energy locations of *DNA* conformations are close to the locations of significant *PoB* within the *y*-range *[GRP_L, GRE_L]*.

Besides, by its terminating role, the stop-codons signal to the ribosome to stop translation. Such operation in the physical language can be described as the rising of an energy barrier preventing further development of energy-consuming process of interest. Then, termination of assembly activity of ribosome can be communicated as the appearance of an energy barrier in the *y*-range *[GRP_L, GRE_L]*. So, in this sense, the linkage stop-codon ↔ energy barrier can be real.

Notice that similar reasonings can apply to characterize an energy position in the *y*-range *[GRP_L, GRE_L]* of the start-codon as well. Then, an appearance of the start-codon may be interpreted as the disappearance of the energy barrier preventing the process of the protein's assembly.

To generalize, we observe that the signal codons, such as the start- and stop-codons, may have an energy dimension, which assumes an energy interconnection between the structure of *GC* and the process of the protein's assembly. Then, it may signify the channel for energy management of the assembly of *GC* nucleotides as of the primary structure of *DNA*, which opens one more connection between *DNA* and *GC*. Unfortunately, the definition of an energy dimension of other *GC* codons is not available as of now, which prevents further development of the above idea.

Meanwhile, turn back to what preceded the announced digression. Accounting for the above-mentioned linkage *PoB* ↔ energy barrier, it is possible to conclude that the found logical chain can be extended further, likewise *PoB* ↔ energy barrier ↔ stop-codon.

Finally, gathering all the above said together, it is possible to build the chain shown in Schema 8.1, implying that denoted energy elements are bound and positioned within the same *y*-range *[GRP_L, GRE_L]*.

It is worth noting that the discovered logic can be applied to explain the fact that not only the position but also the number of occurrences of the known energy elements. It is assumed that the triplicity as a physical phenomenon is observed in the number of the significant *PoB* (3.1.4), the stop-codons (section 3.1.12.2), the triple-based structure of *GC* (3.1.7), and the number of major *DNA*-conformations [46].

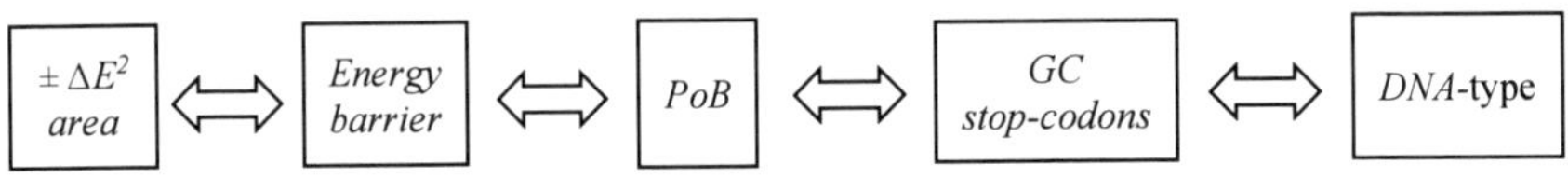

Schema 8.1. Schematic interconnections between key components of the energy landscape in the cell.

In conclusion, pay attention to one more thing. All our energy analysis is based on (2.3), which defines the importance of *n*-structuring. Direct dependence on *n* in the structure of (8.1) evidently points out that the position of energy cylinders follows the *n*-discrete spectrum (2.3). Then, ultimately, the energy cylinders and the *DNA*-conformations $\Delta E^2{}_n$ as their embodiments, come as eigen *n*-harmonics of the energy spectrum in evolution.

Then, it is possible to distinguish the major harmonics corresponding to the basic *A*-, -*B*-, *Z*-conformations from the minor harmonics corresponding to the rare *C*-, *D*-, *E*-, and any other conformations. In this sense, the rare conformations in *CEL* may correspond to the high-order *n*-harmonics with $n \geq 3$ (3.1.4).

3.2.7. *DNA* Nurseries

Take another look at (8.2) to disclose the physical meaning of condition $\Delta E^2 = 0$. From the previous sections, it turns out that within the workable zone, the process of permanent adjustment of *CEL*-solutions through the sign-alternating changes of ΔE^2 is occurring. On the top level, probably, it is done through playing with parameters of the *CEL* energy configurations, such as diameter, height, sense polarity, and so on, which can be treated as optimizing of existing energy guides and adapting them to the real energy environment.

Adapting *DNA* to the changeable energy environment assumes simultaneous transformation of all available *DNA*-conformations (*A*-, *B*-, *Z*-, and existing intermediate ones). In this sense, an emergence of particular conformation is determined by the specific energy conditions in the point of generation.

Energy optimization becomes even more complicated if to account for the fact that, in living cells, it is quite common for *DNA* to not be the whole construction but the composition of the interleaving segments of *A*-, *B*-, and *Z*-type [47]. Another complicative factor here is that the features of conformation are subject to time changes, when the once formed conformation is not forever. It assumes that the *A*-, *B*-, and *Z*-conformations are capable of transforming into each other [48].

Among the above features, possibly, the most critical element is the transition between the helix sense, as it is an energy-consuming process, and the transition from one helix sense to another in terms of energy is not beneficial for the "mature" conformations at all [48].

Accounting for the so varied pattern, it looks reasonable to assume that the specific "mature" conformation reflects the existing energy conditions for some reasonably long time. Then, the energy beneficial way for evolving cell is to stick to a *3*-step conformation plan. In the first step, cell supports the creation of a pool for *DNA* embryos or any other energy-saving *DNA* form, including all *DNA*-conformations. In the second step, the "winner" of this competition in the existing energy environment is defined. In the last step, the "winner" grows up to the "mature" state. Based on the known extraordinary efficiency of all in-cell energy processes, the presence of the in-cell *DNA* embryo's pool, perhaps, is not insensible. Similar mechanism would permit cell to avoid unnecessary energy expenses.

3.2.8. *DNA* Signatures in *CEL* Solution and Hypothesis for Energy Skeleton

In the previous sections we dealt with sufficiently speculative objects in *SEE* space, like energy components $E^T{}_n$, $E^C{}_n$, and E_n, energy flows, and factors, and so far, it has met our purposes. However, in the development of any theory, sooner or later, comes an important moment of "grounding" for the suggested approach, i.e., the binding of scholastic reasonings to the real-world manifestation. That is why below we briefly consider possible output of suggested findings to span the footbridge from the virtual to the real *DNA*'s energy carcass.

The good thing is that actual "grounding" has already started when it was proposed a cooperation between the material objects such as nucleotides and energy phases of *OTS* (section 3.1.7).

Another positive signal is that the physical basis for a more "materialistic" view on things is already intuitively rooted in the *CEL* concept (1.1, 1.2) under the name *CEM* that is applicable for an essentially broader consideration, meaning logical extension on other conserved quantities (*CQ*) like electrical charge, mass, and so on. Dynamics of these *CQ* can be considered separately for each single *CQ* (it could be a subject for another applied research) or as a composite of all contributed items, as it was done in this book.

It should be noticed that construction of a solution based on the *CEM* framework, i.e., the charged nonzero mass centres, immediately lands our academic consideration to the ground, though it raises question on a material carrier of the found energy solution. By a fortuitous combination of circumstances, such material carrier in the form of an energy image has been already presented (8.1). Also, it is worth noting that (8.1) ultimately assumes reconfiguration of the charged mass centres and the spatially-time changes in this arrangement signifies the existence of real energy flows.

In molecular biophysics, the idea that *DNA* can be modelled with reasonable accuracy as an extended polymer (cylinder) was well established by the end of last century [49]. Bearing this in mind, recall that in this chapter, we have discovered the natural appearance of a *3D* energy-related cylinder (8.1) as a result of the intuitive development of *SEE* in phase space. This *CEL* energy cylinder demonstrates its origin in competition between factors of inheritance and variability (section 3.2.3), the phenomenon of energy affinity (section 3.2.2), the mechanism for the formation of complementary antiparallel strands (section 3.2.1), the dynamics of (8.1) as a result of the optimization of an energy exchange (3.2.3), the physical grounds for the interconnection of major energy components (Schema 8.1), the phenomenon of triplicity in the number of stop-codons, significant *PoB*, the energy phases of *SEE*, and *DNA*-conformations (section 3.1.7).

So, to the best of our knowledge, the abovesaid permits put forward the hypothesis that the found *CEL* energy solution (8.1) to the discussed extent is capable of simulating the subset of functional behaviour for the real *DNA* double helix.

3.2.9. Final Remarks to Chapter 3.2

Since the mid-nineteenth century, the Darwin-Wallace mechanism of natural evolution has been the main driving force responsible for the origin of new species, and this view, with some modifications, has survived to the present day [50]. Actually, all major functions of *DNA* can be considered through the lens of evolution, including storing genetic information, directing protein synthesis, determining genetic coding, heredity, and differentiation. To stay on top of its evolutionary mission, *DNA* acquired the well-recognizable physical form and the set of functions. So, the established structure of *DNA* is a nucleic acid double-stranded, antiparallel, right-handed helix, where the strands are connected by hydrogen bonds [51].

In the late 1970s, alternate non-helical models of *DNA* were temporarily considered as a potential solution to issues in *DNA* replication in plasmids and chromatin. However, due to the experimental advances in *X*-ray crystallography, and later the nucleosome core particle, and the discovery of topoisomerases, these models were set aside in favour of the double-helical model [52, 53].

These days, the chemical structure of *DNA* is well-known, and scientific society ultimately came to the conclusion that, unfortunately, the chemistry approach is unable to fully explain the complexity of the *3D* structures of *DNA*. That is why the non-chemical models like *CEL* are currently on the market, meaning their usefulness for understanding *DNA* origin and functioning.

In this context, the ratio (8.1) comes as an energy framework for the construction of *DNA* superhelix. At that, discovered form rather comes up as an energy channel between parts of the energy spectrum of evolution, providing optimal balance between investments into the processes of inheritance and variability. Recall that in the previous chapter the meaning of energy evolution as an optimization process coming through competition between inheritance and variability was disclosed.

A number of researchers believe that the structure of our Universe from the very beginning is based on the principle of modularity [54, 55]. It presumes that even for the most chaotic phenomena, a suitable less or more stable pattern between the key parameters of that phenomena can be identified. The aggregate of these parameters can then be considered as an infrastructure for the studied phenomenon.

The principle of modularity, which generally works for multidimensional entities, allows mapping to space of low dimensionality in the form of convenient spectral discreteness [56]. The latter works for the stochastic processes [57], especially for those that are characterized by an infinite number of dimensions [58].

Though natural evolution takes one of the top places in the list of chaotic phenomena for sure, at the same time it is well recognized that the evolutionary process possesses the structure, the organization, and the set of unique features [59].

In line with the above, highlight the introduction in section 3.2.6 of the physical meaning of energy cylinders (8.1) as n-harmonics of the energy spectrum of *OTS* evolution. Accounting for the close connection of these energy cylinders with *DNA*-conformations, the physical meaning of known *DNA* conformations as the n-spectral nodes with stochastic m-filling deserves to be studied in detail.

In this sense, pay attention to the possible energy position of major *DNA*-conformations with regards to the position of *PoB*. So, the *LSP* ($n = 2$) is considered to be the most stable type (Chapter 2.2), which matches the probabilities of energy fluxes in and out of the *OTS*; then it is consistent to think that in *PoB* terms the *LSP* may correspond to the *B*-conformation. Developing this line of thinking, it is possible to suggest that the less stable *A*-conformation can be located in the vicinity of the outstanding *GRP*, where the most drastic change of *OTS* parameters is observed. Finally, the different and odd-looking *Z*- conformation may be found close to an energy position of *GRE* point, which interfaces the areas of highly stochastic behavior (Figure 8.6).

Conclusion of Chapter 3.2

1. Segregation of *SEE* space creates *2* originally independent sets of energy phases (strands) populated with the speckles of nucleotides, the left one $[GRP_L, SP]$ and the right one $[SP, GRP_R]$.
2. Energy affinity between the strands: (a) makes strands mutually dependent; (b) stabilizes energy structure of strands; (c) creates a twisting moment that contributes to the formation of a *3-D* strand structure. Solution (8.1) comes as an energy framework for *DNA* superhelix as it combines construction of primary, secondary, and tertiary structure of *DNA*, where connection $GC \leftrightarrow DNA$ is supported by the *GC* codon matrix as the primary structure of *DNA*.
3. The result of complete optimization of *SEE* comes as a cylindrical energy form that appears as an energy exchange channel between the

processes of inheritance and variability. A geometrically perfect cylinder corresponds to stationary energy exchange, while the deformation of a cylinder testifies to non-zero energy exchange. In this sense, *DNA* can be thought of as a *SEE*-carrier structure of *GC* within the chain $SEE \leftrightarrow GC \leftrightarrow DNA$.

4. *DNA* conformations are eigen *n*-harmonics of the energy spectrum of evolution.
5. An energy exchange process in a cell is possible in the (GRP_L, GRP_R) range only.
6. The formation of the stable critical elements of energy infrastructure in cell, such as *PoB*, energy barriers, sign-alternating $\pm\Delta E^2$ area, *DNA* conformations, *GC* signal codons, is permitted within the range $(GRP_L, GRE_L]$ only.
7. The items presented above permit putting forward the hypothesis that the *CEL* energy cylinder, to the extent discussed, is capable of simulating the subset of energy behaviour for the real *DNA* double helix.

Conclusion

This book considers a stochastic energy exchange (*SEE*) in an open system as an ordinary physical phenomenon, and originally, in Part 1, it does not mention the application of this theory to evolutionary processes. Things change in Part 2, where, through force of circumstances, the author has to interpret obtained findings in the light of evolutionary changes. Finally, Part 3 has a quite evolutionary orientation. The point is that the phenomenon of evolution in this book arises from the deep physical analysis of the foundations of a quite usual and ubiquitous natural event of energy exchange. And this is what differenties the suggested theory from the theories that were developed especially to simulate the process of evolution, the origin of *GC*, and *DNA*.

It is quite possible that if we take not the *SEE* but something else having the same or close mathematical structure, for example, continuous flow of finances through existing banking infrastructure [60], ultimately, we may come to the same general conclusion for the presence of evolutionary regularities, of course, maybe hidden under the umbrella of multiple details.

The core of *SEE* theory is an analytical model of the energy development of *OTS* based on the concept of an infinite number of conserved energy links with a thermal bath. We knowingly established an infinite number of energy links, i.e., energy sources and energy sinks, considering that such conditions are critical in the evolutionary process. It was done because, in the author's opinion, the number of various energy connections between an object and an external world is what makes evolving entities different from the stagnating ones. Hence, we think that for evolving entities, the number of above links should be incommensurably higher than for non-evolving ones. As a consequence, we replaced the classic *ECE* (1.1) with the system (1.2) and, finally, (1.5), which permits finding of solutions in quadratures.

The author believes that approximation of infiniteness of energy links finds reasonable support in data science theory [61] that declares that for higher accuracy of result in the mass processes, when each single element is a product of random choice, the broad ranging of data, even doubtful, takes critical importance to complete the whole picture. In this sense, the suggested concept of *CEL* is based on as many data patterns as possible due to an assumption of links infiniteness. That is why, from a theoretical standpoint, the author assumes that *CEL* approximates the reality of energy exchange with sufficient accuracy.

However, an infinite number of energy links is not capable of providing an efficient way for the transfer of inheritance information. That is why, to survive and move forward, *OTS* has to reduce an infinitude of links to the finite set of highly protected units, which implants protection of inheritance information directly into the structure of *SEE*. Doing this critical step, nature rejected the infiniteness and continuity in favour of the finiteness and discreteness.

Notice that structurally, the *CEL* approach features a fundamental relation with the provisions of Hamiltonian mechanics, Euclidean geometry, and thermodynamics within the united framework of the energy conservation law, which manifests in the phenomenon of energy development through the mechanism of time translation symmetry breaking. That is why physical, geometrical, and linear algebra methods, tied by parameter k (4.3), are directly applicable for cross-description and interpretation of obtained results.

Another major result is that the suggested concept finds intuitive manifestation in the Darwin-Wallace theory of natural selection, thereby providing the high-level look at the composition of *GC* and *DNA*. Then, we observe the way to merge the biological and physical sciences and place the biological sciences within a more general physicomathematical framework. Hence, in considering approach, the phenomenon of natural evolution comes as a means for implementing conservation laws through the optimization of regularities of energy exchange. As mathematically the same form is applicable not only for energy but also for other conserved quantities, it testifies that support for evolutionary changes is deeply within the foundation of nature's law. Consequently, as conservatism is a form for the appearance of conserved laws and time homogeneity, then, in other words, an evolution is a means for the time existence of *OTS*.

Notice that (1.1) has definite physical limits for its validity due to the form of gravitational curvature. It means that Einstein's general theory of relativity connects gravity to the geometry of spacetime and particularly to its curvature, which is a derivative of the gravitational mass. So, accounting for the absence of significant centres of mass, such as black holes, and that the utilized coordinate system, generally, presumes the finite physical dimensions, it is thought that (1.1, 1.2) is valid across space objects as big as the solar system or less.

The presence of diverse phases and stages of evolution permits us to talk about the existence of an energy infrastructure in the course of evolution. A major contribution to the formation of energy infrastructure is the discreteness of the energy spectrum (2.3), called (n-infrastructure). Besides, it is revealed that there exists a second level of discreteness. It emerges due to the competition between factors of inheritance and variability (6.5) in the m-infrastructure. The driving force for the occurrence of m-infrastructure is natural optimization in evolution between the factors of inheritance and variability. It is worth noticing that optimization fluctuations of local inheritance-variability ratio serve as an additional resource of evolution at the stage of the discrete spectrum.

The author sees the major advantage of the presented theory in a high consistency of material starting from an energy conservation law in the $\{y, \Upsilon\}$ form (1.1) and completing, ultimately, by the same law in the $\{\Delta E^2, \eta\}$ form (8.2). It confirms the fact that the built theoretical construction is completely compatible with existing physical paradigms. Also, it indicates the directional level for analysis of an energy structure of *GC* and *DNA* through interconnection $SEE \leftrightarrow GC \leftrightarrow DNA$.

Conducted analysis demonstrates that the phenomenon of energy evolution comes as a whole complex of mutually connected features. The important result is that the absence of any of these features, even looking sufficiently insignificant, means a paralysis of all others and interruption or even termination of the evolutionary process. It suggests that it is not possible to evolve in a feature-by-feature way. An evolution either comes in all varieties of its features or does not come at all.

Finally, in this research, an evolution process does not show total commitment to the casual way of analysis of minimizing and maximizing the key quantities. Of course, the finding of extremal patterns still was an important part of the research. However, the most critical points in the energy evolution of *OTS* rather point out the influence of other processes; it is an optimization or harmonizing by throwing off the dynamic balance nodes in all or the majority of involved parameters. That is exactly what a real "magic wand" of evolution is!

References

Introduction

[1] Glancy, J., Stone J. V., Wilson S. P. 2016. How self-organization can guide evolution. *R. Soc. Open Sci.* 3: 160553.

[2] *The Free Dictionary by Farlex*, https://www.thefreedictionary.com/evolution

[3] Rybczyk, J. A., *Time and Energy, The Relationship Between Time, Acceleration, and Velocity and its Affect on Energy* 2001. http://www.mrelativity.net/TimeEnergy/TimeEnergy.htm.

[4] Aryan, S. R., The Relationship B/W Time and Energy, *International Journal of Advancements in Research and Technology*, vol. 1, no. 5, p. 126-136, 2012.

[5] Turing A. M. The Chemical Basis of Morphogenesis. \\ *Philosophical Transactions of the Royal Society of London. Series B, Biological Sciences.* — 1952. — Vol. 237. — pp. 37-72.

[6] Newman, S. A., 2003, "*From Physics to Development: The Evolution of Morphogenetic Mechanisms*," in Müller and Newman 2003: 221–239 (chap. 13). doi:10.7551/mitpress/5182.003.0019.

[7] Quispel, A. Pre-biological evolution. *Acta Biotheor* 18, 291–315 (1968). https://doi.org/10.1007/BF01556732.

[8] Jaeger, J. and J. Sharpe, 2014, "*On the Concept of Mechanism in Development*," in Minelli and and Pradeu 2014a: 56–78 (chap. 4).

[9] Pratt AJ. Prebiological evolution and the metabolic origins of life. *Artif Life*. 2011 Summer;17(3):203-17. doi: 10.1162/artl_a_00032. Epub 2011 May 9. PMID: 21554111.

[10] Einstein, A. (1916), *Relativity: The Special and the General Theory*, Berlin, *ISBN* 978-3-528-06059-6.

[11] Luisi, P. L. 2014. The Minimal Autopoietic Unit. *Orig Life Evol Biosph.* 44(4):335-38.

[12] Muller A. W., Schulze-Makuch D. *Thermal Energy and the Origin of Life Origins of Life*. 2006. Vol. 36. Iss. 2. P. 177-189.

[13] Skinner, J. E., Molnar, M., Vybiral, T., Mitra, M. Application of chaos theory to biology and medicine. *Integr Physiol Behav Sci.* 1992 Jan-Mar;27(1):39-53. doi: 10.1007/BF02691091. PMID: 1576087.

[14] Noble D. Evolution viewed from physics, physiology and medicine. *Interface Focus. 7,* 2017. 20160159, http://doi.org/10.1098/rsfs.2016.0159.

[15] Fox R. F. *Energy and the Evolution of Life. — 1988.* — Oxford: W. H. Freeman & Co. — p. 182.

[16] Penrose R. *The Road to Reality. — 2004.* — New York: Alfred A. Knopf. — p. 1136.

[17] Feistel R., Ebeling W. Evolut*ion of Complex Systems, Selforganisation, Entropy and Development.* — 1989. — Berlin: Springer. — p. 242.

[18] Sousa T., Mota R., Domingos T., et al. Thermodynamics of organisms in the context of dynamic energy budget theory \\ *Phys. Rev. E.* – 2006. – Vol 74. - p. 051901.

[19] Aoki I. *Entropy Principle for the Development of Complex Biotic Systems: Organisms.* — 2012. — Amsterdam: Elsevier. — 122 p.

[20] Zhu Y. Large-*scale Inhomogeneous Thermodynamics: and Applications for Atmospheric Energetics.* — 2003. — Cambridge: Cambridge International Science Publishing. — 620 p.

[21] Phillips D. J. *Information Dynamics in Complex Systems: The Evolution of an Infinite Network.* — 2016. — Seattle: Amazon Digital Services LLC. — p. 77.

[22] Lubbock, A. L. R., Carlos, F. L., Programmatic modeling for biological systems, *Current Opinion in Systems Biology*, Vol. 27, 2021, 100343, ISSN 2452-3100.

[23] Hsieh J. S. *Principles of Thermodynamics.* — 1975. — Washington, D. C.: Scripta Book Company. — pp. 800.

[24] Britannica, The Editors of Encyclopaedia. "Conservation Law." *Encyclopedia Britannica.* — 26 Apr. 2021. — https://www.britannica.com/science/conservation-law.

[25] 25. Cartwright, J. H., Giannerini, S., González, D. L. DNA as information: at the crossroads between biology, mathematics, physics and chemistry. *Philos Trans A Math Phys Eng Sci.* 2016, 13;374(2063):20150071. doi: 10.1098/rsta.2015.0071.

Part 1

[1] Bouchiat C., Gibbons G. W. Non-integrable quantum phase in the evolution of a spin-1 system: a physical consequence of the non-trivial topology of the quantum state-space \\ Journal de Physique — 1988. — Vol. 49. — Iss. 2. — pp. 187 – 199.

[2] Sobolev S. L. Convolution of functions. \\ in Hazewinkel M. Encyclopedia of Mathematics, Springer. 1994. — https://encyclopediaofmath.org/wiki/Convolution_of_functions.

[3] Roy R., Oliver F. W. J. Elementary functions. Lambert W function. in NIST Handbook of Mathematical Functions. ed. by F. W. J. Oliver, D. W. Lozier, R. F. Boisvert, C.W. Clark. — 2010. — New York: NIST & Cambridge University Press. — p. 127.

[4] Haken H. Synergetics, *an Introduction: Nonequilibrium Phase Transitions and Self-Organization in Physics, Chemistry, and* Biology, 3rd ed. — 1983. — New York: Springer-Verlag. — p. 355.

[5] Prigogine I. From Poincaré's divergences to quantum mechanics with broken time symmetry. \\ *Zeitschrift für Naturforschung* — 1997. — Vol. 52a — pp. 37-47.

[6] Maturana H., Varela F. *The Tree of knowledge. — 1992.* — Boulder, Shambhala Publications Inc. — 269.

[7] Morozov, A. N., Skripkin, A. V. Linear integral transformations for non-Markovian random processesлинейных интегральных преобразований для описания немарковских случайных процессов \\ *Russian researches*. 2007. Vol. 119. 1243-1251.

[8] Gibbs J. W. *Elementary Principles in Statistical Mechanics.* — 1960. — New York: Dover Publications. — p. 207.

[9] Benguigui L. The different paths to entropy — arXiv.org Solid State Institute and Physics department Technion — Israel Institute of Technology 32000 Haifa. — Israel — 32 pp.

[10] Levin B. P. *Theoretical basics of statistical radio engineering. — 1989.* — M: Radio and communication. — p. 656.

[11] Wheeler N., *Simplified production of Dirac Delta Function Identities.* Reed College Physics Department, November 1997. — pp. 1-18.

[12] Protter M. H., Morrey C. B., Jr. Differentiation under the Integral Sign. *Intermediate Calculus.* 2nd ed. — 1985. — New York: Springer. — p. 655.

[13] Marsh C. *Introduction to Continuous Entropy. — 2013.* — Department of Computer Science, Princeton University. — pp. 1-13.

[14] Park S. Y, Bera A. K. Maximum entropy autoregressive conditional heteroskedasticity model. \\ *Journal of Econometrics* (Elsevier) — 2009. — Vol. 150. — Iss. 2. — pp. 219–230.

[15] Amann H., Arendt W., Neubrander F., et al. *Functional Analysis and Evolution Equations: The Gunter Lumer Volume.* — 2008. — Basel: Birkhauser. — MR 240215. — 637 p.

[16] Swenson R. Emergent attractors and the law of maximum entropy production: Foundations to a theory of general evolution. \\ *Sys. Res.* — 1989. — Vol. 6. — Iss. 3. — pp. 187-197.

[17] Moldavanov A. Abiogenesis-Biogenesis Transition in Evolutionary Cybernetic System \\ *Cybern Syst,* doi: 10.1080/01969722.2021.1991662, https://www.tandfonline.com/doi/full/10.1080/01969722.2021.1991662.

[18] Shpenkov, G. P. *Physical meaning of imaginary unit i.* — 2013. — p. 9. — https://shpenkov.com/ImaginUnitRus.pdf.

[19] Roubicek T. Chap. 17 in: *Mathematical tools for Physicists.* (Ed. M. Grinfeld) — 2014. — Weinheim: J. Wiley. — pp. 551-588., doi: 10.1080/01969722.2021.1991662.

[20] Bohr N. Collected Works. *Foundations of Quantum Physics* I ed. by J. Kalckar. — 1985. — Amsterdam: North-Holland. — Vol. 6. — pp. 316–330, 376–377.

[21] Gingrich T. R., Horowitz J. M., Perunov N. at al. Dissipation Bounds All Steady-State Current Fluctuations \\ *Phys. Rev. Lett.* — 2016. — Vol. 116. — Iss. 12. — 120601.

[22] Pietzonka P., Barato A., Seifert U. Universal bounds on current fluctuations \\ *Phys. Rev. E* — 2016. — Vol. 93. — 052145.

[23] MacKay, D. J. C. *Information Theory, Inference, and Learning Algorithms,* Cambridge University Press, Cambridge, 2003.

[24] Hsieh J. S. *Principles of Thermodynamics.* — 1975. — Washington, D. C.: Scripta Book Company. — pp. 800.

[25] Mathematical Calsulus, Ilyin, V. A., Sadovnichi, V. A., Sendov, B. H., part 1, ed. 3 Tikhonov A. N. 2004. M.: Prospect. 662 pp.

[26] Moldavanov, A. "Theoretical Aspects of Radiative Energy Transport for Nanoscale System: Thermodynamic Uncertainty," *J. Comput. Theor. Trans, 50(3)*, 236-248, 2021.

[27] Horowitz, J. M., T. R. Gingrich, "Thermodynamic uncertainty relations constrain non-equilibrium fluctuations," *Nat. Phys. 16,* 15–20, 2020.

[28] Laurendeau, N. M. *Statistical Thermodynamics: Fundamentals and Applications.* —2005. — New York: Cambridge University Press. — p. 466.

[29] Nalewajski R. On entropy/information continuity in molecular electronic states. \\ *Molecular Physics.* — 2016. — Vol. 114. — No 1. — pp. 1225-1235.

[30] Lindhard J. Complementarity' between energy and temperature. In *The Lesson of Quantum Theory.* ed. by de Boer J., Dal E., Ulfbeck O. — Amsterdam: North Holland Publishing — 1986. — p. 99-112.

[31] Faigon A. Uncertainty and Information in Classical Mechanics Formulation. *Common Ground for Thermodynamics and Quantum Mechanics.* 2003, Arxiv:Quant-Ph/0311153.

[32] Moldavanov, Andrei, Signatures of Substorm Onset in Thermodynamic Model of Geomagnetic Tail. *Adv. Space Res.* Vol. XXX, 2024, XXXXXX, ISSN XXXX-XXXX, https://doi.org/10.1016/j.jastp.2024.XXXXXX.

[33] Gibbs, J. W., *Elementary Principles in Statistical Mechanics.* New York: Dover Publications, 1960.

[34] Bazarov I. P. *Термодинамика, 5-е изд.* — 2010. — СПб.—М.—Краснодар: Лань. — 384 с. (in Russian).

[35] Laurendeau, N. M. *Statistical Thermodynamics: Fundamentals and Applications.* —2005. — New York: Cambridge University Press. — p. 466.

[36] Huang K. *Introduction to Statistical Physics*, 2nd ed. — 2009. — Boca Raton: Chapman and Hall/CRC. — p. 334.

[37] Siegmund, M., Pankratov, O. Violation of the continuity equation in the Krieger-Li-Iafrate approximation for current-density functional theory. \\ *Phys. Rev. B.* — 2011. — 83. — 045113.

[38] Blum H. F. *Time's Arrow and Evolution, 1st ed.* — 1951. — Princeton: Princeton University Press. — p. 252.

[39] Serway, R. A., Jewett, J. W., *Peroomian, V. Physics for scientists and engineers with modern physics, 9th ed.* — 2013. — Pacific Grove: Brooks Cole. — pp. 1616.

[40] Park S. Y, Bera A. K. Maximum entropy autoregressive conditional heteroskedasticity model. \\ *Journal of Econometrics* (Elsevier) — 2009. — Vol. 150. — Iss. 2. — pp. 219–230.

[41] England J.L. Dissipative adaptation in driven self-assembly. \\ *Nature Nanotechnology.* — 2015. — Vol. 10. — pp. 919-923.

Part 2

[1] Livio M. The Golden Ratio: The Story of Phi, *The World's Most Astonishing Number.* — 2002. — New York: Broadway Books. — p. 294.

[2] de Vries M. *The Whole Elephant Revealed: Insights Into the Existence and Operation of Universal Laws and the Golden Ratio.* — 2012. — John Hunt Publishing. — p. 420.

[3] Andrei Moldavanov. Signatures of Substorm Onset in Thermodynamic Model of Geomagnetic Tail. *Adv. Space Res.* Vol. XXX, 2024, XXXXXX, ISSN XXXX-XXXX, https://doi.org/10.1016/j.jastp.2024.XXXXXX.

[4] Porazik, P., Johnson, J.R., Kaganovich I., Sanchez E. Modification of the loss cone for energetic particles \\ *Geophys. Res. Lett.* — 2014. — Vol. 41. — Iss. 22. — pp. 8107-8113.

[5] Andrei Moldavanov. Classical Right-Angle Triangles and Golden Ratio. \\ *Forum Geometricorum.* — 2017. — Vol. 17 — pp. 433—437.

[6] Haken H. Synergetics, *an Introduction: Nonequilibrium Phase Transitions and Self-Organization in Physics, Chemistry, and Biology*, 3rd ed. — 1983. — New York: Springer-Verlag. — p. 355.

[7] England J. L. Dissipative adaptation in driven self-assembly. \\ *Nature Nanotechnology.* — 2015. — Vol. 10. — pp. 919-923.

[8] Mora T., Bialek W. Are Biological Systems Poised at Criticality? \\ *J. Stat. Phys.* — 2011. — Vol. 144. — pp. 268-302.

[9] Santos, F. A. N., Raposo E. P., Coutinho-Filho M. D., et al. Topological phase transitions in functional brain networks. \\ *Phys. Rev. — 2019. — Vol. E 100.* — 032414.

[10] Vaikuntanathan S., Gingrich T. R., Geissler P. L. Dynamic phase transitions in simple driven kinetic networks. \\ *Phys. Rev.* — 2014. — Vol. E 89. — 062108.

[11] Bruss F. T. Sum the odds to one and stop \\ The Annals of Probability. *Institute of Mathematical Statistics.* — 2000. — Vol. 28. — Iss. 1. — pp. 1384–1391.

[12] Russell K. G. Estimating the Value of e by Simulation \\ *The American Statistician.* — 1991. — Vol. 45. — Iss. 1. — pp. 66–68.

[13] Andrei Moldavanov. Theoretical Aspects of Radiative Energy Transport for Nanoscale System: Thermodynamic Uncertainty \\ J. *Comput. Theor. Trans.* — 2021. — Vol. 50. — Iss. 3. — pp. 236-248.

[14] Moldavanov, A., *BioSystems* (2022) doi:10.1016/j.biosystems.2022.104607.

[15] Beekman, A. J., Rademaker, L. van Wezel, J. *An Introduction to Spontaneous Symmetry Breaking.* — 2019. — arXiv:1909.018202.

[16] Quantum Technology News. Google May Have Created an Unruly New State of Matter: Time Crystals. *Popular Mechanics.* — 2021.

[17] D.V. Else, C. Monroe, C. Nayak, N. Y. Yao, *Annu. Rev. Condens. Matter Phys. 11*, 467 (2020).

[18] Baianu, I. C. On Asymmetry in Biology and Nature. \\ *Nature Proceedings.* — 2012. — 3. — p. 1-4.

[19] Moldavanov A. Abiogenesis-Biogenesis Transition in Evolutionary Cybernetic System \\ *Cybern Syst*, doi: 10.1080/01969722.2021.1991662.

[20] Andrei Moldavanov. Signatures of Substorm Onset in Thermodynamic Model of Geomagnetic Tail. *Adv. Space Res.* Vol. XXX, 2024, XXXXXX, ISSN XXXX-XXXX, https://doi.org/10.1016/j.jastp.2024.XXXXXX.

[21] Evans, Merran, Nicholas Hastings, and Brian Peacock. *Statistical Distributions.* 2nd ed. New York: J. Wiley, 1993.

[22] Andrei Moldavanov. Functioning of Autonomous Financial Center Under Permanent Bidirectional Capital Flow from Unlimited Number of Terminals \\

System Analysis & Mathematical Modeling. — 2020. — Vol. 2. — Iss. 4. — pp. 5-18.

[23] Falkovich, Gregory (2018). *Fluid Mechanics.* Cambridge University Press. ISBN 978-1-107-12956-6.

[24] N.A. Campbell, J.B. Reece, R. Heyden. *Biology.* London: Pearson. 2004.

[25] Quantum Technology News. Google May Have Created an Unruly New State of Matter: Time Crystals. *Popular Mechanics.* — 2021, https://www.insidequantumtechnology.com/news-archive/popular-mechanics-google-may-have-created-an-unruly-new-state-of-matter-with-time-crystals. Accessed 5 Aug 2021.

Part 3

[1] *Molecular Biology of the Cell,* 4th edition. Bruce Alberts, Alexander Johnson, Julian Lewis, Martin Raff, Keith Roberts, and Peter Walter. New York: Garland Science; 2002.

[2] *Cell Energy and Cell Functions. Essentials of Cell Biology, Unit 1.3, Cell Biology for Seminars,* Unit 1.3. eBooks. https://www.nature.com/scitable/topicpage/cell-energy-and-cell-functions-14024533/.

[3] *Organic Chemistry,* 10 ed., John McMurry, last updated: Aug 05, 2024. https://openstax.org/books/organic-chemistry/pages/29-7-the-citric-acid-cycle.

[4] Alberts B, Johnson A, Lewis J, Raff M, Roberts K, Walter P. Energy conversion: Mitochondria and chloroplasts. In: Gibbs S, editor. *The molecular biology of the cell.* 4th edition. New York: Garland Science; 2002. p. 769.

[5] https://www.sciencedaily.com/releases/2018/03/180306093304.htm.

[6] https://zeenews.india.com/science/ancient-microbes-produced-oxygen-earlier-than-thought-2087791.html.

[7] Novozhilov AS, Wolf YI, Koonin EV. Evolution of the genetic code: partial optimization of a random code for robustness to translation error in a rugged fitness landscape. *Biol Direct.* 2007 Oct 23;2:24. doi: 10.1186/1745-6150-2-24.

[8] Saito, H. The RNA world 'hypothesis.' *Nat Rev Mol Cell Biol* 23, 582 (2022). https://doi.org/10.1038/s41580-022-00514-6.

[9] Kondratyeva, L. G., Dyachkova, M. S. & Galchenko, A. V. The Origin of Genetic Code and Translation in the Framework of Current Concepts on the Origin of Life. *Biochemistry Moscow* 87, 150–169 (2022). https://doi.org/10.1134/S0006297922020079.

[10] Wolf YI, Koonin EV. On the origin of the translation system and the genetic code in the RNA world by means of natural selection, exaptation, and subfunctionalization. *Biol Direct.* 2007 May 31;2:14. doi: 10.1186/1745-6150-2-14.

[11] Freeland SJ, Knight RD, Landweber LF, Hurst LD. Early fixation of an optimal genetic code. *Mol Biol Evol.* 2000 Apr;17(4):511-8. doi: 10.1093/oxfordjournals.molbev.a026331.

[12] Stuart, A. Harrison, Raquel Nunes Palmeira, Aaron Halpern, Nick Lane. A biophysical basis for the emergence of the genetic code in protocells, *Biochimica et Biophysica Acta - Bioenergetics*, 1863, 2022, 148597, https://doi.org/10.1016/j.bbabio.2022.148597.

[13] Crick, F. H. C. The origin of the genetic code, *Journal of Molecular Biology*, Vol. 38, 1968, 367-79, https://doi.org/10.1016/0022-2836(68)90392-6,

[14] Aldana M, Cazarez-Bush F, Cocho G, Martnez-Mekler G. Primordial synthesis machines and the origin of the genetic code. *Physica A.* 1998;257(1):119–127.

[15] Gusev VA, Schulze-Makuch D. Genetic code: Lucky chance or fundamental law of nature? *Physics of Life Reviews.* 2004;1(3):202–229.

[16] Koonin, E. V, Novozhilov A. S. Origin and evolution of the genetic code: the universal enigma. *PMID*: 19117371, PMCID: PMC3293468, doi: 10.1002/iub.146 (2009).

[17] Xie, P. 2017. The origin of genetic codes: from energy transformation to informatization. *Biodiv. Sci.* 25(1): 94-106.

[18] Paul, N., G. F. Joyce, 2004. Minimal self-replicating systems, *Current Opinion in Chemical Biology*, Volume 8, Issue 6, 634-639. https://doi.org/10.1016/j.cbpa.2004.09.005.

[19] Strick TR, Allemand JF, Bensimon D, Bensimon A, Croquette V. The elasticity of a single supercoiled DNA molecule. *Science.* 1996 Mar 29;271(5257):1835-7. doi: 10.1126/science.271.5257.1835.

[20] Ganti T. *The Principles of Life/* 2003. Oxford University Press, Oxford. 201 pp.

[21] Moldavanov, Andrei. Energy Infrastructure of Evolution for System with Infinite Number of Links with Environment \\ *BioSystems* — 2022 — Vol. 213. — 104607 — doi: 10.1016/j.biosystems.2022.104607.

[22] Feinberg, M. (1963). "Fibonacci-Tribonacci." *Fibonacci Quarterly.* 1: 71–74.

[23] Shu, Jian-Jun. 2017. "A new integrated symmetrical table for genetic codes." *BioSystems.* 151: 21—26.

[24] Ryser, Herbert John (1963), Combinatorial Mathematics, *The Carus Mathematical Monographs 14*, Mathematical Association of America.

[25] Yoshimura T., Esaki N. Amino acid racemases: Functions and mechanisms. *J. Biosci. Bioeng.* 2003;96:103–109. doi: 10.1016/S1389-1723(03)90111-3.

[26] Cooper GM. *The Cell: A Molecular Approach.* 2nd edition. Sunderland (MA): Sinauer Associates; 2000. The Origin and Evolution of Cells. Available from: https://www.ncbi.nlm.nih.gov/books/NBK9841/.

[27] https://en.wikipedia.org/wiki/DNA_and_RNA_codon_tables.

[28] https://openoregon.pressbooks.pub/mhccbiology102/chapter/the-genetic-code/.

[29] Feric, M., Misteli, T. Trends in cell biology. *Phase separation in genome organization across evolution.* Vol. 31, Iss. 8, 671-685. 2021.

[30] Sood, V., Misteli, T. The stochastic nature of genome organization and function, *Current Opinion in Genetics & Development*, Volume 72, 2022, Pages 45-52.

[31] Wang, Y., Huabin, Zhou, X. S., Huang, Q., Li, S., Zhirong, L. L., Luhua, C., Lai Z. Charge Segregation in the Intrinsically Disordered Region Governs VRN1 and DNA Liquid-like Phase Separation Robustness, *Journal of Molecular Biology*, Volume 433, Issue 22, 2021, 167269.

[32] Kauffman, S. A. in *The Origin of Order: Self-Organization and Selection in Evolution* (Oxford University Press, Oxford, 1993).

[33] Skinner, J. E., Molnar, M., Vybiral, T., Mitra, M. Application of chaos theory to biology and medicine. *Integr Physiol Behav Sci.* 1992 Jan-Mar;27(1):39-53. doi: 10.1007/BF02691091.

[34] Scharf, Y. A chaotic outlook on biological systems, *Chaos, Solitons & Fractals*, Vol. 95, 2017, pp. 42-47, https://doi.org/10.1016/j.chaos.201 6.12.013.

[35] Freeland, S. 2013. The Evolutionary Origins of Genetic Information. *BioLogos.* Online publication.

[36] Völker, J. 2020. Genetic code evolution and Darwin's evolution theory should consider DNA an 'Energy Code.' *Rutgers.* Online publication.

[37] Klump, H., Völker, J., & Breslauer, K. (2020). Energy mapping of the genetic code and genomic domains: Implications for code evolution and molecular Darwinism. *Quarterly Reviews of Biophysics*, 53, E11. doi:10.1017/S0033583520000098.

[38] Ursache, R., Nieminen, K., Helariutta, Y. (2013). Genetic and hormonal regulation of cambial development. *Physiol. Plant.* 147 36–45. 10.1111/j.1399-3054.2012.01627.x.

[39] Sonneborn, T. M. (1965). Degeneracy of the genetic code: extent, nature, and genetic implications. In *Evolving genes and proteins*. ed. Bryson, V. and Vogel, H. New York: Academic Press. pp. 377–397.

[40] Cobb M, Comfort N (April 2023). "What Rosalind Franklin truly contributed to the discovery of DNA's structure." *Nature*. 616 (7958): 657–660.

[41] James D. Watson, *The Double Helix: A Personal Account of the Discovery of the Structure of DNA* (1968), Atheneum, 1980.

[42] Bansal M (2003). "DNA structure: Revisiting the Watson-Crick double helix." *Current Science*. 85 (11): 1556–1563.

[43] Cell Energy and Cell Functions. Essentials of Cell Biology, Unit 1.3, *Cell Biology for Seminars*, Unit 1.3. eBooks. https://www.nature.com/scitable/topicpage/cell-energy-and-cell-functions-14024533/.

[44] *Organic Chemistry*, 10 ed., John McMurry, last updated: Aug 05, 2024. https://openstax.org/books/organic-chemistry/pages/29-7-the-citric-acid-cycle.

[45] Cooper GM. The Cell: A Molecular Approach. 2nd edition. Sunderland (MA): Sinauer Associates; 2000. The Origin and Evolution of Cells. Available from: https://www.ncbi.nlm.nih.gov/books/NBK9841/.

[46] https://bio.libretexts.org/Bookshelves/Microbiology/Microbiology_(OpenStax)/11%3A_Mechanisms_of_Microbial_Genetics/11.04%3A_Protein_Synthesis_(Translation)].

[47] *Organic Chemistry, 10 ed.*, John McMurry, last updated: Aug 05, 2024. https://openstax.org/books/organic-chemistry/pages/29-7-the-citric-acid-cycle.

[48] Jurečka, P., Zgarbová, M., Černý, F., & Salomon, J. (2024). Multistate B- to A-transition in protein-DNA Binding – How well is it described by current AMBER force fields? *Journal of Biomolecular Structure and Dynamics*, 1–11. https://doi.org/10.1080/07391102.2024.2327539.

[49] Yoshimura T., Esaki N. Amino acid racemases: Functions and mechanisms. *J. Biosci. Bioeng.* 2003;96:103–109. doi: 10.1016/S1389-1723(03)90111-3.

[50] Kutschera, U., Niklas, K. J. The modern theory of biological evolution: an expanded synthesis. *Naturwissenschaften* 91, 255–276 (2004). https://doi.org/10.1007/s00114-004-0515-y.

[51] Wager, John. (2023). New Perspective on Hydrogen Bonding. *ACS Omega*. 8. 10.1021/acsomega.3c05838.

[52] Lev I. Verkhovsky, 2020. DNA: the Double Helix or the Ribbon Helix? Physics of Biology. viXra.org. *Physics of Biology*. https://vixra.org/abs/1803.0104.

[53] *Proc. Natl. Acad. Sci. USA Vol. 75,* No. 9, pp. 4092-4096, September 1978 Biochemistry Some implications of an alternative structure for DNA (DNA structure/kinky helix/chromatin structure/DNA replication). V. SASISEKHARAN, N. PATTABIRAMAN, AND GOUTAM GUPTA.

[54] Callebaut, W., D. Rasskin-Gutman. (Eds.), *Modularity: understanding the development and evolution of natural complex systems.* (MIT Press, Cambridge, 2005), pp. 207-19.

[55] Simon, H. A. *Models of Discovery*. (D.Reidel, Dordrecht, 1977), 10.1007/978-94-010-9521-1, pp. 475.

[56] Rayner, A. D. M. *Degrees of Freedom: Living in Dynamic Boundaries.* (ICP, London, 1997), pp. 328.

[57] Steven, L. Rev. *Financ. Stud.* 6(2), 327-343, (1993).

[58] Steinerberger, S. *Acta Math Hung* 130, 321–339 (2011).

[59] Glancy, J., Stone JV, Wilson SP. 2016 How self-organization can guide evolution. *R. Soc. Open Sci.*3: 160553.

[60] Shreve, E., Steven (2004). *Stochastic calculus for finance*. New York: Springer. ISBN 9780387401003. OCLC 53289874.

[61] Aurelien Geron. Hands-On Machine Learning with Scikit-Learn, Keras, and TensorFlow. 2019. O'Reilly Media.

[62] Doig, A. J. Improving the Efficiency of the Genetic Code by Varying the Codon Length—The Perfect Genetic Code, *Journal of Theoretical Biology*, Vol. 188, Issue 3, 1997, pp. 355-360, ISSN 0022-5193.

[63] Naranan S., Balasubrahmanyan, Information theory and algorithmic complexity: applications to linguistic discourses and dna sequences as complex systems. Efficiency of the genetic code of DNA. *J. Quant. Linguistics*. Vol. 7, Iss. 2. pp. 129-151, 2010.

[64] Moldavanov, A. (2021). Abiogenesis-Biogenesis Transition in Evolutionary Cybernetic System. *Cybernetics and Systems*, *53*(7), 607–614. https://doi.org/10.1080/01969722.2021.1991662.

Appendices

Appendix A. Geometry of Evolutionary Triangles

A.1. Relations in the Right-Angle Triangle

All relations below assume the equivalence of area S and energy E.

It is proved that if in arbitrary right ΔAB_nC_n (Figure A.1) legs' ratio $k = 2^{\pm 1}$, then holds

$$(\frac{2E^T{}_n}{E^C{}_n})^p = (\frac{E^F{}_n}{2E^T{}_n})^p = \varphi^p \tag{A.1}$$

and ratio $2E^T/E^C$ matches the number of golden ratio φ, where $p = \pm 1$.

Afterwards, this theorem is generalized for the case of arbitrary rational k

$$(\frac{2E^T{}_n}{E^C{}_n})^p = \gamma(k)(\frac{E^F{}_n}{2E^T{}_n})^p \tag{A.2}$$

where $\gamma(k)=(4/k^2)^p$.

Removing γ from (A.2), obtain

$$E_n{}^C(E_n{}^C + 2E_n{}^T) = (kE_n{}^T)^2 \tag{A.3}$$

A.2. *K*-Basis Formalism

Mathematical formalisms developed above use dependence on variables y (B_nC_n) and k (Figure A.1). As within the considering approach, y and k are bound, then it is convenient to rid off y and keep dependence on k only. It means that in k-basis it is assumed that $B_nC_n = 1$, $A_nC_n = k$, and $A_nB_n = \sqrt{k^2+1}$, where $n = 1, 2, \ldots$ (Figure A.1).

Also, energy in k-basis

$$E_n = \frac{k_n}{2} \tag{A.4}$$

From Figure A.1, parameter k

$$k = tg\angle\ ABC = \frac{BC}{AC} \tag{A.5}$$

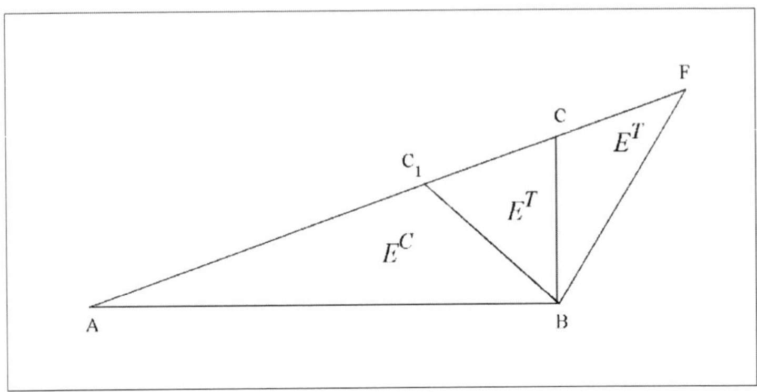

Figure A.1. Triangle ABF with area $E^F = E^C + 2E^T$. A major part of ΔABF is right-angled ΔABC. Triangles CC_1B and CFB are isometric.

A.3. Basic Formulas

Write down some formulas we will use further:

$$f_k = \frac{\sqrt{k^2+1}+1}{k} \tag{A.6}$$

$$E^C{}_n = \frac{B_nC_n{}^2}{2}\left(k - \sqrt{1 - \frac{1}{k^2+1}}\right) \tag{A.7}$$

$$E^T{}_n = \frac{B_nC_n{}^2}{2}\sqrt{1 - \frac{1}{k^2+1}} \tag{A.8}$$

$$E^F{}_n = \frac{B_nC_n{}^2}{2}\left(k + \sqrt{1 - \frac{1}{k^2+1}}\right) \tag{A.9}$$

where f_k is similarity ratio.

A.4. Theorem on "Collapse" of Random Secants

Actually, (A.2) deals with the family of secants within an angle composed by the tangents as shown in Figure A.2. Which secant at the given tangent will be implemented is a matter of random choice. However, at fixed *k*, choice is predefined due to (A.2), and the only one case is realized for one secant with a fixed ratio of its external and internal parts.

Consider the more general case when (A.2) is valid, but *AC1n* and *AFn* are arbitrary (Figure A.1). Register *n* and move to *k*-basis. For brevity, denote $AC1n = \tilde{L}_e$, $C_{1n}F_n = \tilde{L}_i$, and accounting that $ACn = k$, $BnCn = BnDn = 1$, then rewrite (A.2) for $p = 1$ as

$$\tilde{L}_e(\tilde{L}_e + \tilde{L}_i) = k^2 \tag{A.10}$$

where $\tilde{L}_e$ is the random external part of secant, $\tilde{L}_i$ is the random internal part of the secant within circle B_n, denote $\tilde{L}_e + \tilde{L}_i$ as L_S.

So, we have two mutually dependent random quantities, $\tilde{L}_e$ and $\tilde{L}_i$, each with its own support range. At that, as follows from Figure A.2, random selection of L_S reduces to the agreed change $\tilde{L}_e$ and $\tilde{L}_i$, when the set of far points for intersection L_S with circle *Bn* belongs to arc *DnFnCn*, while the set of near points to arc *DnC1nCn*.

Prove the following theorem.

Theorem 3-7. Given the length of the tangent to the circle, the support range for random changes of the secant is equal to the maximum of its internal part and matches the diameter of the circle, while the deterministic form of (A.2) *corresponds to the secant through centre of the circle.*

Proof. At random selection of secant, the far point of intersection of L_S and circle *Bn* slips on the arc *DnFnCn*, and random length L_i in projection on line *AFn* varies within $[0, L_i] = 2R$, where *R* is radius *Bn*.

On the other hand, the nearest point of intersection between L_S and circle *Bn* slips on the arc *DnC1nCn*, and random length L_e in projection on the line *AFn* changes in the range $[L_e, L_e + R]$.

Combining the last two statements, obtain that the support range in projection on line *AFn* changes within the range $[0, L_i] = 2R$, i.e., within the diameter of circle *Bn*.

As $L_S = \sqrt{k^2+1} + 1$, $L_e = \sqrt{k^2+1} - 1$, then $L_i = L_S - L_e = 2$, i.e., equals to diameter in k-basis, which proves the theorem.

So, random variations of the secants L_S can be reduced to a deterministic change of L_e and a random change of L_i. At that, the range of random changes "collapses" to the range $C1nFn$, numerically equal to the diameter of the circle as shown in Figure A.2, A.3.

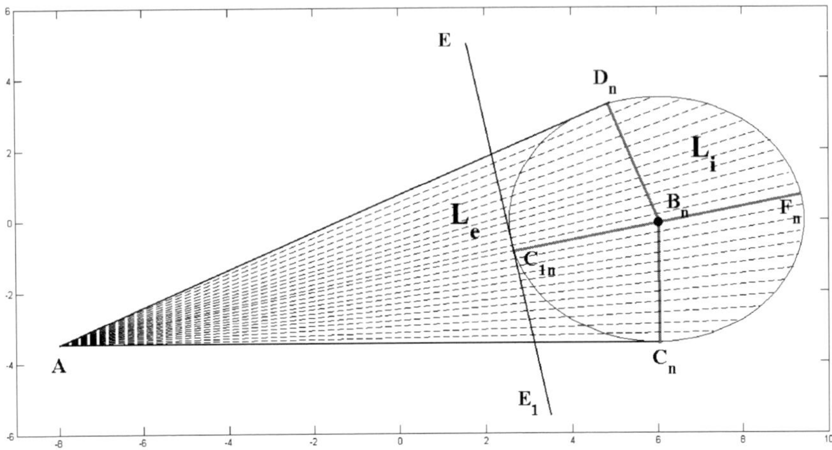

Figure A.2. Family of random secants from point A at given tangent kS_n^T (AD_n and AC_n). Random secants fill up an angle D_nAC_n continuously. Secant L_S is a sum of the random external L_e and random internal L_i parts. Swing for random changes of L_S is the maximum of its internal part L_i and matches the diameter of circle B_n. Line EE_1 provides a reference for changes of secants.

A.5. Theorem on Ratio of Areas E^C and E^T for Any k

Consider a more general case when (A.3) holds but E^T and E^C are arbitrary. From here, there exists an infinite number of embodiments for ratio E^C: E^T (obviously that for fixed n, quantity $E^C + E^T = E = const$ at any E^C and E^T), at which E^T alternates from zero (secant becomes tangent) to maximum $2E^T$ (secant crosses center of circle B). In line with this, E^C alternates from minimum E^C_{min} (secant through center O) to the maximum kE^T (secant becomes tangent). So, E^C in (A.3) changes in the range $[E^C_{min}, kE^T]$), while $2E^T$ in the range $[0, 2E^T]$. Then,

Theorem 3-8. At conditions of theorem 3-7, a support range for the random variations of the internal part of secant $2E^T$ all the time exceeds the corresponding range of external part of secant E^C, i.e., *inequality*

$$2E^T \geq kE^T - E^C \tag{A.11}$$

holds at any *k*.

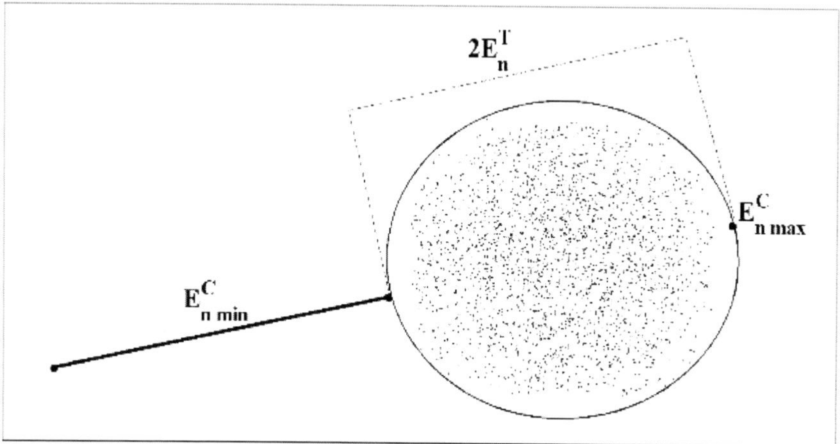

Figure A.3. The same as in Figure A.2 presented in the complex form. The difference is that the range for random $\hat{E}^T$ and $\hat{E}^C$ is shown as a circle with a diameter of $2E^T$. Minimal and maximum possible energy E_{n+1} at given E_n are also shown.

Proof. Based on (A.7) and (A.8), from (A.11)

$$\frac{2k}{\sqrt{k^2+1}} \geq k\left(\frac{k+1-\sqrt{k^2+1}}{\sqrt{k^2+1}}\right) \tag{A.12}$$

Transforming (A.12), obtain

$$\sqrt{k^2+1} \geq k-1 \tag{A.13}$$

From (A.13) follows that

$$k \geq 0 \tag{A.14}$$

which holds at any k.

So, within the discussed framework, an energy E^C demonstrates less changeability compared to energy E^T. Actually, even

$$E^T \geq kE^T - E^C \tag{A.15}$$

is also valid.

Due to $E^{tr} = E^C + E^T$, the quantities E^C and E^T are dependent. Nevertheless, for convenience, we further use them as the self-dependent factors. Emphasize that the segregation of E^C and E^T is defined by the division of the triangle and does not have any connection to the features of E^C and E^T themselves. The procedure of division for an area of a triangle on E^C and E^T (fine structure of area) will be called a partitioning (decomposition) of the triangle's area.

A.6. Eigenvalues for Operator of *OTS* Evolution

In the y-interval [GRP_L, GRP_R], the resultant action of operator L on flow y is equivalent to a change in value and direction of y. In terms of k-basis, such transformation is equivalent to the change in form and size of a rotating triangle (Figure A.4).

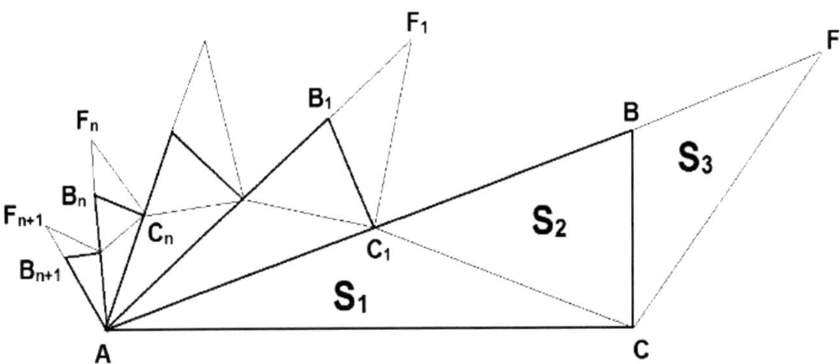

Figure A.4. Consecutive evolutionary changes in rotating right angle triangle in k-базисе.

A.7. Maximizing Support Range for Fluctuations of Secant

Find the position of the secant that corresponds to the maximum ratio of the internal part of secant to its total length. With this purpose, prove the following theorem.

Theorem 3-9. On assumption of a unform probability distribution, the maximum ratio of the internal part of secant to its total length is observed if the secant crosses the center of the circle.

So, it is proved that the maximum ratio of the internal part of the secant to its total length is observed if the secant crosses the center of the circle.

Appendix B. Selection of Combinatorial Method

As the population of energy phases is done on a random basis, the classic recipe here is to apply the apparatus of mathematical combinatorics.

Let me summarize what is known about methods of combinatorial calculations. In the context considered, the proper combinatorial methods are variation, permutation, and combinations.

Combination is the selection of objects, with or without repetition, where the order does not matter. As the existence of order is not important, combination is hardly applicable for our purposes in ordered codon arrangement.

The next method is permutation. Permutation is the arrangement of objects (with or without repetition), where order does matter. Generally, permutation could be used for our purposes with the exception that it works not with selections of objects but with objects themselves, which makes permutation a less general method.

Instead, variation is the arrangement of selections of objects (with or without repetition), where the order of the selected objects matters.

So, it looks like that variation is an optimal way to calculate the number of possible selection variants in placement of nucleotides against the energy phases and account order of constructed codons, whereas combination with repetition can be used in case order does not matter.

Appendix C. Hypothetical Realizations of *GC*

C.1. Realization 64^1

This realization reduces to the formal placement of *64* nucleotides over the original energy stage *[0, GRP_R]*. Obviously, it simply copies *64* nucleotides to the original energy range *[0, GRP_R]*, which yields *64 x 1 = 64* outcomes and an appropriate number of combinations

$$C_1(64) = 64 \tag{C.1}$$

As much as this realization is efficient in terms of the number of combinatorial outcomes, it looks nonsensical from the inheritance standpoint by eliminating any diversity.

Drawbacks of this realization are obvious: (a) over-expenditure of nucleotides for coding of inheritance information; (b) non-zero risk of wrong identification for nucleotides with too close energy signatures; (c) from (C.1), only *1* amino acid can be coded by this realization, which obviously is not sufficient for proper transfer of inheritance information (Figure C.1).

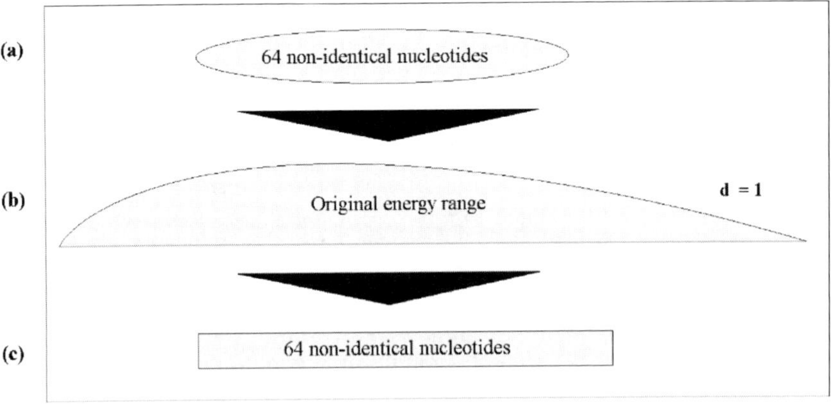

Figure C.1. Schematic view for draw in placement of *64* nucleotides split up over *1* energy phase. Dependence $\Upsilon(y)$ is shown in the middle panel in light-grey, while possible outcome is in the lower panel of the chart.

C.2. Realization 8^2

It assumes the placement of eight distinct nucleotides $\{A, B, C, D, E, F, G, H\}$ over two energy phases. One phase (call it *L1*) is the *y*-range $[0, GRP_L]$, and another (call it *L2*) could be the *y*-range $[GRP_L, GRP_R]$. The resulting set comes as *64* doublets randomly populated from the set $\{L1, L2\}$ (Figure C.2), which yields $64 \times 2 = 128$ outcomes, and

$$C_2(8) = 36 \tag{C.2}$$

The drawbacks of this realization are: (a) the risk of taking an unstructured *SEE* scenario if it falls into the *y*-range $[0, GRP_L]$, which is evaluated by the ratio $L1/L2 = 1/(e^2\text{-}1) \approx 14\%$, where *L1* is the *y*-length of the range $[0, GRP_L]$, and *L2* is the *y*-length of the range $[0, GRP_R]$; (b) though it is possible to code *36* amino acids by *64* nucleotides, such codification does not look optimal.

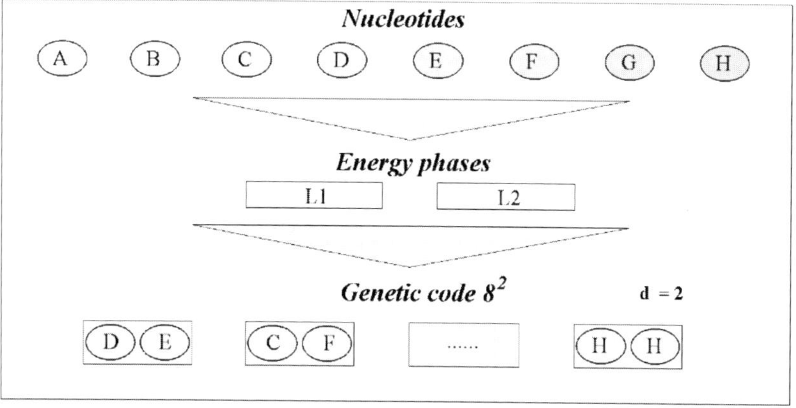

Figure C.2. Schematic view for draw in placement of *8* distinct nucleotides $\{A, B, C, D, E, F, G, H\}$ split up over *2* available energy phases, *L1* and *L2*.

C.3. Realization 2^6

It assumes the placement of two distinct nucleotides over six energy phases. Schematically, the result of such intercoupling is the chart of *64* sextuplets randomly populated from the *2* nucleotide exemplar set $S = \{Y, Z\}$, as shown in Figure C.3, where *Y* and *Z* are nucleotides. It yields $64 \times 6 = 384$ outcomes, and

$$C_6(2) = 7 \qquad \text{(C.3)}$$

Drawbacks: (a) the existence of this realization leads to just *1*-strand *DNA*, i.e., *RNA* form, which rather refers to an archaic realization of *GC* (section 3.1.10). It follows from the well-known mechanism of the secondary structure forming for realization 4^3, when the two strands of *DNA* in a double helix are bonded together by hydrogen bonds between the two pairs of nucleotides on the opposite strands, *G – C*, and *A – U*. However, in the realization 2^6, by definition, there is only one pair of nucleotides, which makes the stability of the double helix questionable; (b) the double triplet structure contradicts the existing single triplet-based structure of modern *DNA*; (c) as per (C.3), the calculated number of amino acids is *7*. Though it is possible to code *7* amino acids by *64* nucleotides, the assembly of proteins from the *7* acids hardly provides a desirable level of diversity.

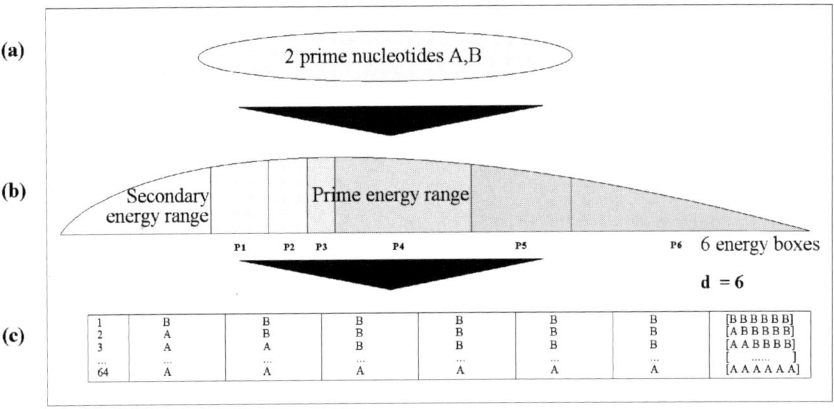

Figure C.3. Schematic view for draw in placement of *2* distinct nucleotides {*A, B*} split up over *6* available energy phases, *Pi1 – Pi6*. Dependence *Y (y)* is shown in the middle panel, while possible outcomes are summarized in the lower panel of the chart.

C.4. *GC* Realization 4^3

It assumes the placement of four nucleotides over three energy phases, and it is a major realization for nucleotide dissemination as it works in modern *DNA*, yielding $64 \times 3 = 192$ outcomes, and

$$C_4(3) = 20 \tag{C.4}$$

Schematically, the result of placement is shown in Figure C.4, where *A, B, C,* and *D* are nucleotides.

Drawbacks: realization 4^3 with *192* outcomes is less efficient than 8^2 with *128* outcomes and 64^1 with *64* outcomes.

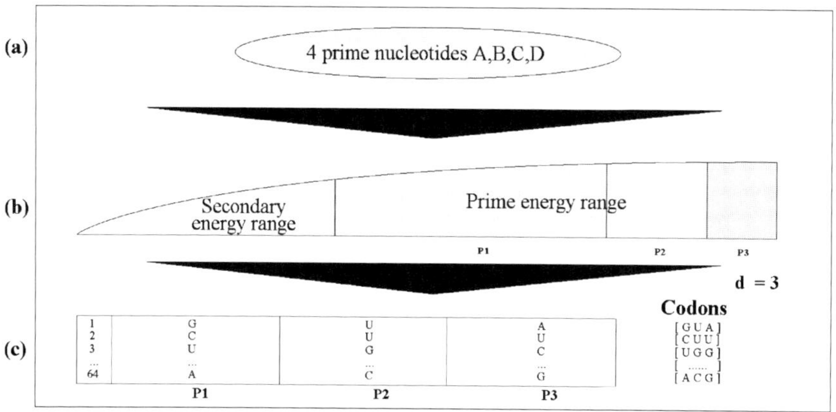

Figure C.4. Schematic view for draw in placement of *4* distinct nucleotides (*A, C, G, T*) split up over *3* available energy phases, *Pi1 – Pi3*. Dependence $\Upsilon(y)$ is partly (for $GRP_L \leq y \leq SP$) shown in the middle panel, while possible outcomes are summarized in the lower panel of the chart.

So, the presented combinatorial approach to *GC* formation yields the degenerative mathematical structure of *GC* with the well-recognizable characteristic numbers *1, 3, 4, 20,* and *64*. The realization 4^3 possesses the least number of drawbacks compared to the other realizations.

Appendix D. η-Representation of Dynamic Balance Points

It is important to highlight that in addition to the existence of the points of dynamic balance at the *y*-level in areas *[GRP$_L$,GRE$_L$]* and *[GRE$_R$,GRP$_R$]* (3.1.4), the phenomenon of dynamic balance also manifests at the η-level.

Indeed, from (6.3)

$$\eta_1(k=2) = -2\varphi,\ \eta_2(k=2) = 2\varphi - 2 = \frac{2}{\varphi} \tag{D.1.a}$$

$$\eta_1(k=\frac{4}{3}) = -1 - \frac{1}{3}(\varphi + \frac{1}{\varphi})^2,\ \eta_2(k=\frac{4}{3}) = -1 + \frac{1}{3}(\varphi + \frac{1}{\varphi})^2 \tag{D.1.b}$$

where φ is the number of golden ratios.

In other words, similarly to *y*-representation, along the η-axis we observe the existence of two points of dynamic energy balance, *GRP* and *GRE*, separated by the point *LSP*

$$\eta_1(k=\frac{3}{2}) = -1 - \frac{\sqrt{13}}{2},\ \eta_2(k=\frac{3}{2}) = -1 + \frac{\sqrt{13}}{2} \tag{D.1.c}$$

Pay attention that *LSP* does not reference φ, so it is not a point of dynamic balance; instead, it is a point of the static energy balance.

Index

U

V